ISOFAR International Society of Organic Agriculture Research

Organic Agriculture in the Tropics and Subtropics

Edited by

Ulrich Köpke

SCIENTIFIC PUBLISHERS (INDIA)
www.scientificpub.com

Print: 2021

Reprint in India, 2010

Published by:
Scientific Publishers (India)
5-A, New Pali Road, P.O. Box 91
Jodhpur - 342 001

E-mail: info@scientificpub.com
www.scientificpub.com

ISBN: 978-81-7233-660-0

Printed at Hinglaj Offset Printers, Jodhpur (India)

Contents

Foreword

Dear reader,

The first volume of ISOFAR's planned scientific series on Organic Agriculture in the tropics and subtropics is lying in front of you. Why this series? Compared with the activities on Organic Agriculture in Europe and North America, and above all its scientific statement and support, even its scientifically based development is still backward in the southern hemisphere. This is regrettable in view of the fact that the research centres of CGIAR and ACIAR were established especially in the South. Thus, it is my personal desire for us to be successful in gaining more support on the part of the centres in the form of personnel and infrastructure for the scientifically based development of Organic Agriculture.

The first volume of our scientific series is versatilely composed. There is a good reason for this. With the first volume we would like to point out the different areas of agronomic research and consultancy in the tropics and subtropics and to stimulate the reader to become self-active in research and teaching in Organic Agriculture. The main target audience of our series are scientists and teachers who, compared with their counterparts in temperate climates, have few research studies available to relate their own research and teaching to.

The next publications in our series will be thematic volumes. Currently, volumes are projected for soil fertility-building concepts, nutrient management, paddy rice production, cotton, coffee, etc. Maybe you, dear reader, will feel inspired to contribute to the further development of Organic Agriculture, especially in the tropics and subtropics, with an article of your own in an upcoming volume.

The editor is grateful to the authors who contributed to this first volume. He is greatly indebted to Mrs Annette Frankenberg for her tireless and diligent efforts as editorial assistant. May this book find a good reception among its readers.

Ulrich Köpke
President ISOFAR

Optimizing the use of BNF in Organic Agriculture – advantages of the tropics

Maria Cristina Prata Neves, José Antonio Azevedo Espindola, José Guilherme Marinho Guerra, Helvécio De-Polli

Embrapa Agrobiologia (www.cnpab.embrapa.br), P.O. Box 74505, Seropédica, RJ, 23890-000, Brazil

1.1 Introduction

Among all plant nutrients, nitrogen (N) is the one required in higher amount for plants and animals. Despite being the most abundant in the terrestrial atmosphere its deficiency in agricultural soils is very common in the tropics. This happens because the combined forms of N are highly soluble, easily leaching and running off the radicular zone or can suffer volatilization and denitrification, particularly intense in tropical soil, returning to the atmosphere. On the other hand, atmospheric N can be supplied to plants through a biological pathway. This unique process is called biological nitrogen fixation (BNF).

In the tropics, animal manure is used as a source of N for crop production in Organic Farming. In some regions, it may be scarce or very expensive. Biologically nitrogen fixation than becomes a source of N for crop production of utmost importance. Besides presenting a minor energy cost in relation to the industrial nitrogen fertiliser, it represents an alternative to make a viable organic production system (Espindola et al., 2004). Optimising its use is a challenge to the sustainability of tropical Organic Farming systems.

1.2 Nitrogen fixing organisms

Several nitrogen fixing organisms have been discovered, mostly from tropical soils (Table 1). The ability of N fixation is restricted to the diazotrophic group of prokaryotic microorganisms (Bacteria and Archae). Some superior plants (mainly plants of the Leguminosae and Gramineae families, and other plant with less significant occurrence in agriculture) when associated to the diazotrophic micro-organisms receive this beneficial effect of BNF. The quantity of nitrogen fixed varies depending on crop, management, and environmental factors. The legume-rhizobium (*Rhizobium, Bradyrhizobium, Azorhizobium, Sinorhizobium, Allorhizobium, Mezorhizobium, Betarhizobium*) symbiosis has been the major biological nitrogen fixing system with practical use in agriculture and the amount fixed is highly significant (Table 2).

Being a biological process, the BNF has a complex regulatory mechanism. With the exception of its success in conventional soybean crop in Brazil (Alves et al., 2002), the BNF process is under-explored in modern conventional agricultural systems that emphasises short cycle crops, without the necessary time for the establishment of the BNF apparatus. In contraposition, Organic Agriculture presents particularly a favourable opportunity to the use of BNF as a strategy for the management of nitrogen in agroecosystems.

Table 1: Diversity of biological nitrogen fixing bacteria associated with legumes and grasses

Host	Diazotrophic bacteria
Leguminosaea (nodules and nodule-like structures)	*Rhizobium, Bradyrhizobium, Azorhizobium, Sinorhizobium, Metarhizobium, Allorhizobium, Betarhizobium* (Willems, 2006)
Leguminosaea – other symbionts	*Methylobacterium* (Samba et al. 1999; Jourand et al., 2004), *Devosia* (Rivas et al., 2003), *Ochrobactrum* (Trujillo et al., 2005) and *Phyllobacterium* (Valverde et al. 2005), *Burkholderia* (Vandamme et al., 2002; Chen et al., 2005) and *Ralstonia* (Chen et al., 2001)
Gramineae	*Azotobacter paspali, Azospirillum, Herbaspirillum, Bacillus, Gluconacetobacter, Bulkholderia,*

Table 2: BNF associated to legume and grass crops

Culture	% N derived from BNF	N fixed (kg ha^{-1})	Ref.*
Legumes			
Arachis hypogeae	w/i	68-116	a
	22-92	37-206	b
	38-92	20-297	c
Pisum sativum	60-81	133-183	b
Phaseolus vulgaris	w/i **	3-32	d
	0-73	0-125	e
Vigna unguiculata	8-89	32-89	a, b
	9-51	11-201	c
Cicer arietinum	8-82	3-141	b
Cajanus cajan	10-81	7-235	b
	4-88	2-92	c
Lens esculenta	39-87	10-192	b
Glycine max	64-81	85-155	a
	70-72	163–195	f
	w/i	159-450	c
Lupinus luteus	79-88	188-327	g
Grass/cereal			
Saccharum officinarun	39-52	110-175	h
Oryza sativa	w/i	20-30	i

Ref.: (a) Boddey et al. (1990); (b) Peoples et al. (1995); (c) Giller et al. (1997); (d) Duque et al. (1985); (e) Hardarson et al. (1993); (f) Urquiaga & Zapata (2000); (g) Unkovich et al. (1994); (h) Boddey et al. (2001); (i) Boddey et al. (1995).

** w/i – without information

1.3 Forms of BNF utilization in organic production systems

1.3.1 Nitrogen-fixing legume plant as direct protein source

Whenever possible, the introduction of legume cash crop in the organic production system may give some farming advantage. Soybean, chick pea, common bean, cowpea, pigeon pea, lentil, and pea can be introduced as monocrop in rotation or intercropped with other crops, as in the case of a traditional intercropping system of common bean and corn of Brazilian family farmers. The BNF contribution for legume plants depends on the establishment of an efficient nodulation with rhizobium bacteria and of favourable soil and climatic conditions as well as adequate management. The BNF can supply more than 90% of the necessary N for the growth of some plants (as demonstrated for peanut, lupin and soybean) in experimental field conditions, incorporating into the system up to 450kg N ha-1 in the case of the soybean (Table 2).

A recent study has shown evidence of the potential of some cultivars of soybean under organic management for the use as green manure (Padovan et al., 2002) with the great advantage of an easy seed supply. Depending on the sowing time, larger amounts of N derived from biological fixation were observed. Biomass production was also affected by the sowing period, although it did not show strictly relation with greater BNF ability (Table 3).

Table 3: Dry matter and N accumulation derived from BNF in the aerial plant part of some cultivars of soybean at different planting time under organic management.

Soybean cultivar	Aerial plant part					
	Dry matter ($Mg\ ha^{-1}$)			N from BNF ($kg\ ha^{-1}$)		
	Planting time[1]					
	1st	2nd	3rd	1st	2nd	3rd
Taquari	5.85 Ab[2]	7.92 Aa	5.73 Abb	79 Bb	112 Aab	146 Aba
Celeste	5.70 Ab	7.33 Aa	5.49 Abb	78 Bb	109 Aab	146 Aba
Sambaíba	7.21 Aa	7.35 Aa	6.14 Aba	151 Aa	88 Ab	185 Aa
Seridó	6.79 Aa	7.70 Aa	7.02 Aa	129 Aa	90 Aa	110 Ba
Conquista	6.06 Aa	6.68 Aa	4.39 Bb	128 Aa	80 Aa	122 ABa
C.V. (%)		13.8			26.9	

Adapted from Padovan et al. (2002)
[1] Time of planting: 1st, September; 2nd, November; 3rd, January.
[2] Same letter in the same column (capital letters) or in the same line (lower case letters) shows no statistic difference among treatments by Tukey test ($p < 0.05$).

1.3.2 Legume plants as green manure

Green manuring consists of an agricultural practice where some type of plants with high capacity of biomass production is cultivated as monocrop in rotation or intercropped with cash crops of economic interest. The green manure plants can be incorporated into the soil or managed in the form of side strip, live or dead mulch. Such practices have been shown viability in several situations (Araújo & Almeida, 1993; De-Polli et al., 1996; Espindola et al., 1998; Espindola et al., 2005; Guerra et al., 2002; Oliveira, 2001). Leguminosae family plants are the most used for green manure due to its high potential for BNF.

Tropical legume green manure crops nodulate well in many situations with indigenous rhizobium, but local observation of this ability is important. In some regions, inoculation with specific strains of rhizobium may be required to provide for an efficient nodulation otherwise plants will not properly develop and the green manure will give poor results.

a) Green manuring in the form of pre-cropping and intercropping with annual cash crop

Traditionally green manure is planted during the fallow season prior to the main crop. Favourable water supply is required for better plant development and soil coverage. With adequate moisture and temperature the amount of biomass produced is very high adding considerable amount o N to the soil after cutting (Oliveira et al., 2003). Legumes may also be cultivated in periods less favourable to its development, associated with the occurrence of low temperatures and shortage in rain (Espindola et al., 1998). In such situation, legume growth is reduced but plants still contribute to the protection of soil and decrease of weed infestation. Table 4 presents a summary of the contribution of BNF in different species of legume and grass with potential for use as green manure.

Table 4: BNF associated to legume and grass used as green manure, live mulch, and forage

Species	N derived from BNF (%)	N fixed (kg ha^{-1})	Period of measurement (days)	Ref.*
Arachis pintoi	72-87	1-7	84	a
Calopogonium	w/i **	64-182	360	a
Centrosema spp.	82-83	41-280	119- 360	a
Crotalaria juncea	80-96	146-221	102-190	a
Clitoria ternatea	77-81	197-249	190-195	a
Desmodium spp.	77-100	24-380	190-360	a
Indigofera tinctoria	70	79	225	a
Lotus comiculatus	30-85	49-109	360	a
Mimosa pudica	w/i	43-83	w/i	b
Medicago sativa	46-92	90-386	360	a
Mucuna deeringiana	96	70	60-70	a
Pueraria spp.	75-88	9-115	72-199	a
	68-92	136-182	w/i	b
Sesbania spp.	13-94	7-209	45-116	b
Macroptilium atropurpureum cv. Siratro	78-92	15-167	360	a
Stylosanthes spp.	60-95	2-75	63-360	a
	63-97	7-212	s/i	b
Trifolium repens	62-93	54-291	360	a
T. pratense	35-87	69-373	360	a
T. subterraneum	50-93	2-206	360	a
Vicia sativa	75	106	w/i	a
Pennisetum purpureum	w/i	106 –165	210	c
Brachiaria decumbens and B.humidicola		30-40	360	d

*Reference: (a) Peoples et al. (1995); (b) Giller et al. (1997); (c) Urquiaga & Boddey (2001); (d) Boddey & Victoria (1986).

** w/i – without information.

De-Polli and Chada (1989) showed the effect of green manuring on crop rotation with maize. Legumes were planted during the fallow period (mild winter) between maize crops in a low fertility Ultisol (Red Yellow Podzolic soil) of the State of Rio de Janeiro, Brazil. Liming was performed in the whole area. Maize received basal phosphorus and potassium fertilization at seeding time in the furrow and green manures were not fertilized. Green manures were planted at the end of summer period (after maize harvesting time), and cut at flowering stage, at the end of the winter period, approximately two months before a new maize planting time. The results showed that fiber content of the tropical green manure was an important factor to be considered in management: Velvet bean (*Mucuna aterrima*) is a green manure with low fiber content in the aerial tissue showed the same effect on maize yield when incorporated or left as mulch. In contrast sunn hemp (*Crotalaria juncea*) and jack bean (*Canavalia ensiformis*) are plants of high fiber content and incorporation into the soil was a better treatment compared with mulch. Maize yields in plots with green manure were equivalent or even better than N treatments (80 to 100 kg ha-1 of N-urea) besides leaving in the soil a considerable amount of C (Table 5). Araújo and Almeida (1993) in a similar soil also determined a fertilizer nitrogen equivalent (FNE) of 80 kg ha-1 of N, using jack bean as green manure.

The effect of legume population density during pre-cropping growth was evaluated recently by Pereira et al. (2005). They determined the amount of N in the aerial part of Sunn hemp (*Crotalaria juncea*) in different densities of seeding and different plant spacing. The result indicated greater value (183 kg ha^{-1} of N) during the rainy season for the treatment of 40 plants per linear meter and 0.30 m spacing between rows. During the dry season, a maximum value of 140 kg ha^{-1} of N was attained with at lower plant density (30 plants per linear meter and 0.30 m spacing between rows). It is important to remind that such effect vary with legume species. Moreira et al. (2003) found that increasing the plant density of pigeon pea (*Cajanus cajan*) resulted in no significant differences in the amount of accumulated N derived from the BNF.

Another possibility for managing pre-cropping green manure consists on its simultaneous growth with a non-legume cash crop aiming to increase the efficiency of the BNF contribution as reported by Aita and Giacomini (2003). Giller (2001) showed that legume-grass intercropping promotes legume BNF due to the increased soil N uptake by the grass roots with lowered N availability to legume therefore favouring the process of nodulation of the legume plants. Similar results have been found by Perin et al. (2004a). They found that BNF contributes as much as 61% of N in plots with intercropped legume-grass (*Crotalaria juncea/Pennisetum americanum*) compared with only 57% in plots where crotalaria was grown as a single crop.

Table 5: Effect of green manure planted during the fallow season in rotation with maize in a low fertility Brazilian Ultisol (Red Yellow Podzolic soil) (maize yields are average of three cropping seasons)

Treatment during fallow period	Management treatment Incorporated	Mulch
	kg/ha	
Velvet bean	2100 bc[2]	2086 bc
Sunn hemp 1st year, jack bean 2nd and 3rd year	2861 a	1674 cd
Spontaneous vegetation (mainly grass)	1369 d	740 e
Spontaneous vegetation + N[1]	2355 b	2318 b
Spontaneous vegetation reduced by mechanical weeding	887 e	569 e

Adapted from De-Polli and Chada, 1989.

[1]80 kg ha^{-1} of N-urea in the first and second years and 100 kg ha^{-1} in the third year.
[2]Same letter in the same column shows no significant difference among treatments by Duncan test ($p<0.05$).

Intercropping the green manure legume plant with an annual cash crop is a system particularly adapted to small holders since it optimizes the use of natural resources. In tropical region fast mineralization of the vegetal residues can lead to rapid loss of N from soils. For this reason intercropping the green manure is an alternative capable to increase the availability of this nutrient for the main crop. An example of such system was demonstrated by Hodtke et al. (1997). They evaluated N supply for a crop of maize by intercropped cowpeas. The authors show the great potential of this practice, especially when cowpeas were planted in two lines between maize rows: grain yield and leaf N content of maize was significantly increased. The effects were observed as early as eight days after legume cutting.

The management applied to the intercropped legume plants affect the amount of nitrogen supplied to the main crop. Ribas et al. (2003a) studied crotalaria intercropped with okra

(*Abelmoschus esculentus*). When legume plants were cut near ground there was an increase of the okra yield followed by higher amount of N accumulated in the leaves (Table 6). In another experiment crotalaria was pruned at 0.80 m height in alternated lines. Later the pruned lines were cut near ground. As a consequence, okra yield was much higher in this innovative crotalaria pruning-cutting system (Ribas et al., 2003b).

Table 6: Fruit weight, yield and accumulated N uptake of okra (*Abelmoschus esculentus*) in mono-cropping or intercropped with *Crotalaria juncea* in different planting density (2000-2001 season)

Treatment	Okra fruits[1]		
	Single fruit weight (g)	Yield (Mg ha^{-1})	N uptake (kg/ha)
Mono-cropping	14.78 A[4]	27.23 B	144.34 B
Intercropped (2 rows of *Crotalaria*)	14.81 A	30.28 A	166.45 A
Intercropped (3 rows of *Crotalaria*)	14.88 A	30.81 A	166.52 A

Adapted from Ribas et al. (2003 a)

[1]Total of 25 harvesting.

[2]Same letter in the same column shows no significant difference among treatments by Tukey test ($p<0.05$).

In another study, Almeida and Araújo (2000) showed that the simultaneous growth of crotalaria (two or three lines between maize rows) provided significant increase in grain yield when compared with single maize plots in a Red Yellow Podzolic sand-clay soil of Seropédica, Rio de Janeiro. The highest increases of 68, 60 and 46% was attained by treatments: three lines of crotalaria cut and left on the soil surface, three lines of crotalaria cut and incorporated into the soil and two lines of crotalaria cut and left on the soil surface, respectively (Table 7).

Table 7: Dry weight and N content of *Crotalaria juncea* aerial plant part and corn yield

Treatment[1]	Crotalaria dry matter (kg ha^{-1})	N content in *Crotalaria* (%)	N acumulated in *Crotalaria* (kg ha^{-1})	Corn yield (kg ha^{-1})
Corn (mono-crop)	-	-	-	2542.9c
C-2-C	2939.3bc	2.56a	75.3b	3711.7ab
C-2-I	2690.2c	2.60a	69.9b	3002.5b
C-3-C	3211.9ab	2.57a	82.5ab	4267.7a
C-3-I	3402.1ab	2.61a	88.8a	4072.9a
Crotalaria (single)	3545.4a	2.64a	93.6a	-
CV (%)	20.59	9.11	20.18	23.25

Adapted from Almeida & Araújo (2000)

[1]C-2-C corn intercropped with two rows of crotalaria cut for mulch; C-3-C corn intercropped with three rows of crotalaria cut for mulch; C-2-I corn intercropped with two rows of crotalaria cut and incorporated in the soil; C-3-I corn intercropped with three rows of crotalaria cut and incorporated in the soil

[2]Same letter in the same column shows no significant difference among treatments by Duncan test ($p<0.05$)

b) Green manuring as permanent live mulching for perennial crops

The benefits of green manures observed for annual crops may also be extended to orchards. Perennial herbaceous legume can be cultivated between rows of trees or other perennial species, creating permanent live mulch for soil protection. The main advantage of this type of green manuring is related to the fact that there is no need of new annual sowing as long as plants develop well enough to support periodical pruning. Several legume species can be used with this purpose, such as forage peanut (*Arachis pintoi*), tropical kudzu (*Pueraria phaseoloides*), galaxia (*Galactia striata*) and siratro (*Macroptilium atropurpureum*) (Perin et al., 2004b; Espindola et al., 2005). However, to avoid competition with the main tree crop caution should be taking such as choosing non climbing species.

A study conducted in an orchard of banana trees cv. "Nanicão" demonstrated the potential of the use of live perennial mulch associates to this culture (Espindola et al., 2006). Amongst the evaluated legumes, the best plant treatment were forage peanut, tropical kudzu and siratro that allowed for a faster vegetative development of the banana trees when compared to the treatment with spontaneous vegetation mainly of *Panicum maximum* grass. Living mulch of tropical kudzu also resulted in higher amounts of N derived from BNF (305 kg ha-1 of N). All legumes transferred N to the banana trees, with levels varying from 24% (siratro) up to 40% (forage peanut) (Table 8). During the second cycle, green mulch with tropical kudzu and siratro resulted in higher banana yields, while the treatment of the forage peanut was lower (Perin et al., 2002). A high competition for water in the second study was the main reason for the poorer performance of the forage peanut.

Table 8: Proportion (Nfbl) and total N derived from BNF in legumes, and proportion of N in banana tree derived from BNF of legumes (Ndfbl) (accumulated values of three harvesting cut) [1]

Cover crop	Nfbl in legume (%)	N from BNF in legume (kg ha^{-1})	Ndfbl In banana tree (%)
Arachis pintoi	66.9a	201.6b	40.5a
Pueraria phaseoloides	86.2a	305.5a	33.7a
Macroptilium atropurpureum	38.2b	89.3c	24.2a
Spontaneous vegetation	-	-	-

[1]Adapted from Espindola et al. (2006)

Same letter in the same column shows no significant difference among treatments by Scott-Knott test ($p<0.05$).

1.3.3 Legume as dead mulching

Dead mulching is a beneficial cultural practice especially in horticulture. Legume straw shredded and left on soil surface protects against soil erosion releases nutrients as well as maintains adequate root temperature and protection against soil erosion (Espindola et al., 2006) favouring the agronomic performance of the crops. In a study where different plant materials were used as dead mulch for a lettuce seedling bed (Oliveira 2005) legume mulch was more efficient for supplying of N for lettuces, in two consecutive cycles, as compared to grass mulch (Table 9).

Table 9: N content in lettuce with different dead mulch

Dead mulch	N content ----(g kg^{-1})----	
	1st crop	2nd crop
Saccharum officinarum	29.8 b	26.0 b
Bambuza sp.	32.4 b	25.8 b
Pennisetum purpureum	28.2 c	25.2 b
Crotalaria juncea	35.3 a	30.3 a
Eritrina poeppigiana	35.8 a	29.0 a
Gliricidia sepium	38.8 a	29.3 a
Cajanus cajan	32.3 b	29.2 a
Mucuna pruriens	36.5 a	28.0 a
Control	27.8 c	23.7 b

Adapted from Oliveira (2005)

[1]Same letter in the same column shows no significant difference amongst treatments by Scott-Knott test ($p<0.05$).

1.3.4 Legume in agroforest system

Legume tree and other tree species that associates with nitrogen fixing microorganisms assume an important role when it comes to the management of landscape, integrating units of production with forest fragments as well as creating vegetal corridor for connecting these fragments (Franco & Balieiro, 2000). This process offers an ecological shelter for local fauna while supplying N for plant growth (Table 10).

Table 10: BNF associated with nodulated trees and shrubs

Species	N derived from BNF (%)	N fixed (kg ha^{-1})	Period of measurement (days)	Ref. *
Legumes				
Acacia holosericea	30	3-6	180	a
Aeschynomene indica	93-100	75-127	116	a
Calliandra calothyrsus	14-48	11-101	90-180	a
	0-87	11-377	90-360	b
Gliricidia sepium	43-75	86-309	360	a
	26-89	170-204	85-360	b
Leucaena	34-78	98-230	90-180	a
	34-100	76-274	85-360	b
Sesbania rostrata	68-94	70-324	45-65	a
Prosopis glandulosa	2-61	40	w/i **	a
Non-legume				
Casuarina equisetifolia	39-90	9-440	180-360	a

*Ref.: (a) Peoples et al. (1995); (b) Giller et al. (1997)

** w/i – without information.

Legume tree increases the recovery of nutrients that are leached out from the root region of the cultures providing the return of the nutrients to the superficial layer of the soil. This is one of the best examples of nutrients cycling in production systems. The trees also create a favourable microclimate for the plants, protecting the soil from strong rain, improving the capacity of soil water retention. *Sesbania rostrata* is a type of shrub that shows one of the highest growth rates. It is able to accumulate about 5 kg N ha^{-1} day^{-1}, according to data collected by Peoples et al. (1995). Tree legumes such as *Inga, Erythrina* and *Gliricidia* are successful example of a multi-use plant used for shading coffee plants and for living fence posts as well as components of alley cropping system.

Managing tree root distribution provides a possibility for managing soil at depth in a sustainable production system. Virginia and Jarrell (1983) found that the presence of woody legumes, such as mesquite (*Prosopis* spp.), can dramatically alter the spatial distribution of nutrients in soil at depth and increase soil N levels in the Sonoran desert ecosystem. This can result in an overall increase in the fertility of the soil. The use of tree in a complex system such as alley cropping generates residue quality as showing to have a major impact on the dynamics of applied residual N in a Nigerian soil comparing high-quality *Leucaena leucocephala* N residue (C/N ratio 11.3, lignin/N ratio 2.2) with low-quality *Dactyladenia barteri* N residue (C/N ratio 24.6, lignin/N ratio 16.1) (Vanlauwe et al. 1998).

1.3.5 Managing legumes in pastures

Although regarded as very important for the sustainability of Organic Farming in the tropics, integrating plant production with animal production is a challenge. Establishing a good pasture with sufficient plant cover as well as providing fodder during the dry season requires an N supply that have to come mainly from BNF. To maintain legumes in a mixed pasture is very important but demands special management to maintain a sustainable population of legumes. Less palatable legumes can be choose to be used in order to overcome over-grazing that must be avoided to prevent the disappearance of the legume from the pasture.

Studies have shown that the presence of *Desmodium ovalifolium* in a pasture of *Brachiaria humidicola* significantly increases the protein content of the associated grass, and the rate and quantity of N recycled through the litter pathway (Cantarutti et al. 2001). The authors concluded that despite of no major effect in animal weight gain due to this legume, the increase in N recycled through the litter provides an increase in the sustainability of pasture production, especially at the lower stocking rate where proportion of legume was higher during the 9-year period of observation (Tarré et al 2001).

Fodder production in a farm area with no grazing may be a simpler management option to keep a balanced stand ratio of legume and grass with substitution of N fertilization by biological nitrogen fixation (De-Polli et al., 1973) (Table 11).

Table 11: Forage legume intercropped with forage grass (*Pennisetum purpureum*). Adapted from De-Polli et al., 1973

Treatment	Fresh total forage matter yielded in four harvesting
	metric ton ha^{-1}
Grass (control)	68.32 bc[1]
Grass + FYM* 2 dm^3 m^{-1} row	67.62 bc
Grass + 2x40 kg N ha^{-1}	97.37 a
Grass + *Macroptilium atropurpureum*	89.37 ab
Grass + *Centrosema pubescens*	62.12 c
Grass + *Pueraria phaseoloides*	74.68 abc
Grass + *Glycine wightii*	63.18 c
Grass + *Stylosanthes guianensis*	99.30 a

* FYM – farmyard manure

[1]Same letter shows no significant difference among treatments by Tukey test ($p < 0.05$).

Legume trees can also integrate to the pastures. They supply shading for good animal welfare, as well as enriching of soil with fixed N. They can be managed as protein bank, or as living fencing poles separating stands (Maradei & Franco, 2000).

1.4 Organic composting enrichment with legume plants

Legume plant residues can be used in the organic compost preparation. It is an interesting practice for Organic Agriculture, mainly in regions where the availability of animal manure is reduced. Legumes are used as an alternative material for N enrichment for the composting process. Leal (2006) showed the composting efficiency prepared from mixtures of *Crotalaria juncea* with Napier grass (*Pennisetum purpureum*), which can be used as seed bed substrate for horticultural plants as well as for their fertilization after transplanting to the field.

1.5 The role of Gramineae

Great advancement in terms of identification of bacterial taxonomy and measurement of the contribution in terms of BFN was obtained regarding to the capacity of the Gramineae plants to host bacterial colonization able to fix atmospheric nitrogen by the group that started in the 60s with Dr Johanna Döbereiner at what is now called Embrapa Agrobiology Research Center.

Several nitrogen-fixing bacteria with different role have been identified as the rhizospheric type of colonization (*Beijerinckia fluminensis* and *Azotobacter paspali*), associative colonization (*Azospirillum lipoferum, A. brasilense, A. amazonense*) and the endophytic colonization (*Herbaspirillum seropedicae, H. rubrisubalbicans, Gluconacetobacter diazotrophicus, Burkholderia silvatlantica* and *B. tropica*), demonstrating the importance of the endophytic colonization and the role of plant genotype to the relationship plant/bacteria (Baldani & Döbereiner 1995; Baldani & Baldani, 2005). These endophytic diazotrophs have been observed only within the intercellular spaces, vascular tissue, aerenchyma and dead cells of their hosts, not in their living cells (James, 2000). The endophytic type of colonization gives competitive advantage for these bacteria when compared to rhizospheric associations where the environment is much more competitive among different kinds of microorganisms.

Sugar cane is planted in furrows using stem cuttings that naturally carries in its interior part diazotrophic bacteria such as *G. diazotrophicus*. In some situation, micropropagation is used aiming the elimination of plant pathogens. In this case, the endogenous diazotrophic bacteria also can be eliminated. To overcome this problem, strain of diazotrophic bacteria can be inoculated at early stage of seedling preparation to keep the benefit of BNF to the cropping system. There is a method to introduce selected endophytic diazotrophic strains during the acclimatization process developed by Reis et al. (1999). The bacterial infection and colonization of the root tissues and aerial parts may occur by penetrating root tip and lateral roots formed during the rooting process (James et al. 1994, 2001).

Quantification of BNF using isotopic technique and N balance showed fixation related to species of grass such as *Paspalum notatum* (De-Polli et al. 1977), *Brachiaria decumbens*, *B. humidicola* and *Panicum maximum* (Boddey & Victoria 1986; Miranda et al. 1990). The potential of BNF in some varieties of sugar cane and flooded rice can be quite significant (Table 1). Studies have demonstrated that in some efficient varieties of sugar cane with endophytic bacteria-stem association may contribute to 60% of the nitrogen necessities of the plant (Boddey et al. 1995). Recent studies have shown the potential of selected strains of nitrogen fixing bacteria used to inoculating sugar cane micropropagated cuttings. A 30% increase in stem production and increased BNF contribution (equivalent to the recommended annual nitrogen fertilization) was observed (Oliveira et al. 2006) in some sugar cane varieties grown under commercial field conditions in a soil of low fertility (Alfisol) without nitrogen fertilization for three consecutive crops.

The potential of BNF in Gramineae has been demonstrated and can be important for Organic Agriculture but its widespread agronomic application still needs to be developed.

1.6 General Comments

All those practices associated with proper preparation of the soil and the use of live or dead mulch for protection can provide a balanced supplement of N as well as other nutrients which are adequate for the development of the plants. These actions allow the diversification of the cover cropping system. Those strategies are within the fundamental principles of Organic Agriculture given support for its sustainability. Proper technological adaptation can occur in many different ways, depending on the climatic conditions and on the local natural resources, as well as the social-cultural conditions.

Understanding Organic Agriculture as a sustainable agricultural production with focus on local use of resources BNF plays a major role by decreasing producer's dependence for external farm inputs while saving natural and non-renewable resources.

On the other hand, BNF as a biological process is naturally regulated (meaning that BNF is modulated by the availability of soil N). That is, greater BNF contributions are attained in soils with low N available to plants.

The Organic Agriculture offers the opportunity to accept the challenge of maximization of the use of the nitrogen and carbon through biological process available in nature. Both of those chemical elements, which are very important for vital process of biogeochemical cycles, receive special attention in organic management of soil. Those elements are components of vegetal tissues and are part of the huge atmospheric reservoir. They are also beneficiaries with the direct use of solar energy, in the case of photosynthesis and indirectly, in the case of BNF. These elements are responsible for the most important components of the human and animal food (carbohydrates, proteins, vitamins, etc) and of organic substances that keep the soil alive.

Looking at the Organic Agriculture like a way to treat the unit of agriculture production as an 'organism', the biological processes of CO_2 and N_2 fixation are clear opportunities to apply this principle.

References

Aita, C. & Giacomini, S.J. (2003) Decomposição e liberação de nitrogênio de resíduos culturais de plantas de cobertura de solo solteiras e consorciadas. Revista Brasileira de Ciência do Solo, **27**, 601-612

Almeida, D.L. de & Araújo, P.A. de (2000) Produção de grãos de milho em sistema de cultivo consorciado com *Crotalaria juncea* em sistema orgânico de produção.: Embrapa Agrobiologia, Seropédica, Relatório Técnico do Projeto Manejo em Agricultura Orgânica, Subprojeto Sistema Integrado de Produção Agroecológica/SIPA) .

Alves, B.J.R.; Boddey, R.M. & Urquiaga, S. (2002) The success of BNF in soybean in Brazil. Plant Soil, **252**, 1-9

Araújo, A.P. & Almeida, D.L de (1993) Adubação verde associada a fosfato de rocha na cultura do milho. Pesquisa Agropecuária Brasileira, **28**, 245-251

Baldani, J.I. & Baldani, V.L.D. (2005) History on the biological nitrogen fixation research in graminaceous plants: special emphasis on the Brazilian experience. An. Acad. Bras. Cienc., **77**(3), 549-579

Baldani, J.I. & Döbereiner, J. (1995) Recent advances in BNF with non-legumes plants. In: International Symposium on Sustainable Agriculture for the Tropics – The Role of Biological Nitrogen Fixation, Angra dos Reis, 1995. Abstracts... Angra dos Reis: The National Centre for Agrobiology Research (EMBRAPA-CNPAB), 53-54

Boddey, R.M.; Oliveira, O.C.; Urquiaga, S.; Reis, V.M.; Olivares, F.L.; Baldani, V.L.D. & Döbereiner, J. (1995 a) Biological nitrogen fixation associated with sugar cane and rice: contributions and prospects for improvement. Plant Soil **174**, 195-209

Boddey, R.M. & Victoria, R.L. (1986) Estimation of biological nitrogen fixation associated with Brachiaria and Paspalum grasses using 15N labelled organic matter and fertilizer. Plant Soil, **90**, (1), 265-292

Boddey, R.M.; Polidoro, J.C.; Rezende, A.S.; Alves, B.J.R. & Urquiaga, S. (2001) Use of the 15N natural abundance technique for the quantification of the contribution of N2 fixation to grasses and cereals. Australian Journal of Plant Physiology, **28**, (9), 889-895

Chen, W.M.; de Faria, S.M.; Straliotto, R.; Pitard, R.M.; Simoes-Araujo, J.L.; Chou, J.H.; Chou, Y.J.; Barrios, E.; Prescott, A.R.; Elliott, G.N.; Sprent, J.I.; Young, J.P.W. & James, E.K. (2005) Proof that *Burkholderia* strains form effective symbioses with legumes: a study of novel *Mimosa* nodulating strains from South America. Appl. Environ. Microbiol. **71**, 7461-7471

Chen, W.M.; Laevens, S.; Lee, T.M.; Coenye, T.; Vos, P.D.; Mergeay, M. & Vandamme, P. (2001) *Ralstonia taiwanensis* sp. nov., isolated from root nodules of *Mimosa* species and sputum of cystic fibrosis patient. Int. J. Syst. Evol. Microbiol. **51**, 1729 – 1735

Cantarutti, R.B.; Tarré, R., Macedo, R.; Rezende, C. de P.; Pereira, J.M., Ferreira, E.; Alves, B.J.R.; Urquiaga, S. & Boddey, R.M. (2001) Forage legume presence and grazing intensity effect on nitrogen dynamics in *Brachiaria* pastures in the south of Bahia, Brazil. Proceedings of the XIX International Grassland Congress, 2001, 102-104.

De-Polli, H. & Chada, S. de S. (1989) Adubação verde incorporada ou em cobertura na produção do milho em solo de baixo potencial de produtividade. Rev. Bras. Ci. Solo **13**, 287-293

De-Polli, H.; Franco, A.A. & Almeida, D.L. de. (1973) Consorciação do capim elefante (*Pennisetum purpureum*) com cinco leguminosas forrageiras tropicais. Instituto de Pesquisa Agropecuária do Centro-Sul, Rio de Janeiro, Bol. Tec. **104**, 9

De-Polli, H.; Guerra, J.G.M.; Almeida, D.L. de & Franco, A.A. (1996) Adubação verde – parâmetros para avaliação de sua eficiência. Castro Filho, C. & Muzilli, O. (eds.). Manejo Integrado de Solos em Microbacias Hidrográficas. IAPAR, Londrina, 225-242

De-Polli, H.; Matsui, E.; Döbereiner, J. & Salati, E. (1977) Confirmation of nitrogen fixation in two tropical grasses by $^{15}N_2$ incorporation. Soil Biol. Biochem. **9**, 119-123

Duque, F.F.; Neves, M.C.P.; Franco, A.A.; Victoria, R.L. & Boddey, R.M. (1985) The response of field grown *Phaseolus vulgaris* to *Rhizobium* inoculation and the quantification of N_2 fixation using ^{15}N. Plant Soil, **88**, 333-343

Espindola, J.A.; Almeida, D.L.; Guerra, J.G.; Ribeiro, R.D. (2004) Gestão do nitrogênio em sistemas orgânicos de produção através da adubação verde. Ciência & Ambiente, **29**, 123-130

Espindola, J.A.A.; Almeida, D.L.; Guerra, J.G.M.; Silva, E.M.R.; Souza, F.A. (1998) Influência da adubação verde na colonização micorrízica e na produção da batata-doce. Pesquisa Agropecuária Brasileira, **33**, 339-347

Espindola, J.A.A.; Guerra, J.G.M.; Almeida, D.L.; Teixeira, M.G. & Urquiaga, S. (2005) Decomposição e liberação de nutrientes acumulados em leguminosas herbáceas perenes consorciadas com bananeira. Revista Brasileira de Ciência do Solo, **30**, 321-328

Espindola, J.A.A.; Guerra, J.G.M.; Perin, A.; Teixeira, M.G.; Almeida, D.L.; Urquiaga, S. & Busquet, R.N.B. (2006) Bananeiras consorciadas com leguminosas herbáceas perenes utilizadas como coberturas vivas. Pesquisa Agropecuária Brasileira, **41**, 415-420

Franco, A.A. & Baleiro, F.de C. (2000) The role of biological nitrogen fixation in land reclamation, agroecology and sustainability of tropical agriculture. Rocha-Miranda, C.E. (ed.) Transition to Global Sustainability: the Contribution of Brazilian Science. Academia Brasileira de Ciências, Rio de Janeiro, 209-233

Giller, K.E. (2001) Nitrogen fixation in tropical cropping systems. 2. ed. CABI Publishing, Wallingford

Giller, K.E.; Cadisch, G.; Ehaliotis, C.; Adams, E.; Sakala, W.D. & Mafongoya, P.L. (1997) Building soil nitrogen capital in Africa. Buresh, R.J.; Sanchez, P.A. & Calhoun, F. (eds.). Replenishing Soil Fertility in Africa. Soil Science Society of America, Madison, Wisconsin,. 151-192. (SSSA. Special Publication, 51)

Guerra, J.G.M.; De-Polli, H. & Almeida, D.L. de. (2004) Managing carbon and nitrogen in tropical organic farming through green manuring. Badejo, M.A. & Togun, A.O. (eds.) Strategies and tactics of sustainable agriculture in the tropics (STASAT). Ibadan (226p), **2**, 125-140

Hardarson, G.; Bliss, F.A.; Cigales-Rivero, M.R.; Henson, R.A.; Kipe-Nolt, J.A.; Longeri, L.; Manrique, A.; Pena-Cabriales, J.J.; Pereira, P.A.A.; Sanabria, C.A. & Tsai, S.M. (1993) Genotypic variation in biological nitrogen fixation by common bean. Plant Soil, **152**, 59-70

Hodtke, M.; Almeida, D.L.; Köpke, U.; Alves, B.J.R.; Urquiaga, S. & Unkovich, M. (1997) Balanço de nitrogênio em diferentes sistemas de produção orgânica para milho e caupi. In: Congresso Brasileiro de Ciência do Solo, 25, 1997, Rio de Janeiro. Anais... Rio de Janeiro: SBCS, Seção temática 4. CD-ROM

James, E.K. (2000) Nitrogen fixation in endophytic and associative symbiosis. Field Crops Research **65**(2), 197-209

James, E.K.; Olivares, F.L.; Oliveira, A.L.M., Reis, Jr. F.B.; Silva, L.G. & Reis, V.M. (2001) Fürther observations on the interaction between sugar cane and *Gluconacetobacter diazotrophicus* under laboratory and greenhouse conditions. J. Exp. Bot. **52**, 747–760

James, E.K.; Reis, V.M.; Olivares, F.L.; Baldani, J.I. & Döbereiner, J. (1994). Infection of sugarcane by the nitrogen-fixing bacterium *Acetobacter diazotrophicus*. J. Exp. Bot. **45**, 757–766

Jourand, P.; Girand, E.; Bena, G.; Sy, A.; Willems, A.; Gillis, M.; Dreyfus, B. & deLajudie, P. (2004|) *Methylobacterium nodulans* sp. nov., for a group of aerobic, facultatively methylotrophic, legume root-nodule-forming and nitrogen-fixing bacteria. Int. J. Syst. Evol. Microbiol. **54**, 2269-2273

Leal, M.A.A. (2006) Produção e eficiência agronômica de compostos obtidos com palhadas de gramínea e leguminosa para o cultivo de hortaliças orgânicas. Universidade Federal Rural do Rio de Janeiro, Seropédica,. 180 p. (Tese de Doutorado em Agronomia, área de concentração Ciência do Solo)

Maradei, M. & Franco, A.A. (2000) Avaliação de dez espécies do gênero *Erythrina* no Rio de Janeiro para uso como moirão vivo. Agronomia, **34** (1/2), 26-30

Oliveira, F.L. (2001) Manejo Orgânico da Cultura do Repolho (*Brassica oleracea* var. capitata): Adubação Orgânica, Adubação Verde e Consorciação. 103 p. Universidade Federal Rural do Rio de Janeiro, Seropédica (Dissertação de Mestrado em Fitotecnia)

Miranda, C.H.B.; Urquiaga, S. & Boddey, R.M. (1990) Selection of ecotypes of *Panicum maximum* for associated biological nitrogen fixation using the 15N isotope dilution technique. Soil Biol. Biochem. **22**, 657-663

Moreira, V.F.; Pereira, A.J.; Guerra, J.G.M.; Guedes, R.E. & Costa, J.R. (2003) Produção de biomassa de guandu em função de diferentes densidades e espaçamentos entre sulcos de plantio. Embrapa Agrobiologia, Seropédica, 5 p. (Comunicado Técnico, 57)

Oliveira, F.F. (2005) Influência da cobertura morta com palha de leguminosas e gramíneas no desempenho de alface (*Lactuca sativa* L.) sob manejo orgânico. Universidade Federal Rural do Rio de Janeiro, Seropédica,.60 p. (Dissertação de Mestrado em Fitotecnia)

Oliveira, F.L.; Ribas, R.G.T.; Junqueira, R.M.; Padovan, M.P.; Guerra, J.G.M.; Almeida, D.L. & Ribeiro, R.L.D. (2003) Uso do pré-cultivo de *Crotalaria juncea* e de doses crescentes de "cama" de aviário na produção do repolho sob manejo orgânico. Agronomia, **37** (2), 19-24

Oliveira, A.L.M. de; Canuto, E. de Lima; Urquiaga, S.; Reis, V.M. & Baldani, J.I. (2006) Yield of micropropagated sugarcane varieties in different soil types following inoculation with iazotrophic bacteria. Plant Soil **284**, 23–32

Padovan, M.P.; Almeida, D.L.; Guerra, J.G.M.; Ribeiro, R.L.D. & Ndiaye, A. (2002) Avaliação de cultivares de soja, sob manejo orgânico, para fins de adubação verde e produção de grãos. Pesquisa Agropecuária Brasileira, **37**, 1705-1710

Peoples, M.B.; Herridge, D.F. & Ladha, J.K. (1995) Biological nitrogen fixation: an efficient source of nitrogen for sustainable agricultural production? Plant Soil, **174**, 3-28

Pereira, A.J.; Guerra, J.G.M.; Moreira, V.F.; Teixeira, M.G.; Urquiaga, S.; Polidoro, J.C. & Espindola, J.A.A. (2005) Desempenho agronômico de *Crotalaria juncea* em diferentes arranjos populacionais e épocas do ano. Embrapa Agrobiologia, Seropédica, 4 p. (Comunicado Técnico, 82)

Perin, A.; Santos, R.H.S.; Urquiaga, S.; Guerra, J.G.M. & Cecon, P.R. (2004a) Produção de fitomassa, acúmulo de nutrientes e fixação biológica de nitrogênio por adubos verdes em cultivo isolado e consorciado. Pesquisa Agropecuária Brasileira, **39**, 35-40

Perin, A.; Guerra, J.G.M.; Teixeira, M.G. & Zonta, E. (2004b) Cobertura do solo e estoque de nutrientes de duas leguminosas perenes, considerando espaçamentos e densidades de plantio. Revista Brasileira de Ciência do Solo, **28**, 207-213

Perin, A.; Lima, E.A.; Espindola, J.A.A.; Guerra, J.G.M.; Teixeira, M.G. & Busquet, R.N.B. (2002) Contribuição da cobertura viva de solo com leguminosas herbáceas perenes no 2° ciclo de produção de bananeiras cultivar Nanicão. Embrapa Agrobiologia, Seropédica, 6 p. (Comunicado Técnico, 53)

Reis, V.M.; Olivares, F.L.; Oliveira, A.L.M.; Reis, Jr. F.B.; Baldani, J.I. & Döbereiner, J. (1999) Technical approaches to inoculate micropropagated sugar cane plants with *Acetobacter diazotrophicus*. Plant Soil **296**, 205–211

Ribas, R.G.T.; Junqueira, R.M.; Oliveira, F.L.; Guerra, J.G.M.; Almeida, D.L.; Alves, B.J.R. & Ribeiro, R.L.D. (2003a) Desempenho do quiaboiro (*Abelmoschus esculentus*) consorciado com *Crotalaria juncea* sob manejo orgânico. Agronomia, **37** (2), 80-84

Ribas, R.G.T.; Junqueira, R.M.; Oliveira, F.L.; Guerra, J.G.M.; Almeida, D.L. & Ribeiro, R.L. D. (2003b) Manejo da adubação verde com crotalária no consórcio com o quiabeiro sob manejo orgânico. Embrapa Agrobiologia, Seropédica, 4 p. (Comunicado Técnico, 59)

Rivas, R.; Willems, A.; Subbarao, N.S.; Mateos, P.F.; Dazzo, F.B.; Kroppenstedt, R.M.; Martinez-Molina, E.; Gillis, M. & Velazquez, E. (2003) Description of *Devosia neptuniae* sp. nov. that nodulates and fixes nitrogen in symbiosis with *Neptunia natans*, an aquatic legume from India. Syst. Appl. Microbiol. **26**, 47-53

Samba, R.T.; De Lajudie, P.; Gillis, M.; Neyra, M.; Spencer-Baretto, M.M. & Dreyfus, B. (1999) Diversity of rhizobia nodulating *Crotalaria* spp. from Senegal, Symbiosis 27, 259–268

Tarré, R.; Macedo, R.; Cantarutti, R.B.; Rezende, C. de P.; Pereira, J.M.; Ferreira, E.; Alves, B.J.R.; Urquiaga, S. & Boddey, R.M. (2001) The effect of the presence of a forage legume on nitrogen and carbon levels in soils under *Brachiaria* pastures in the Atlantic Forest region of the South of Bahia, Brazil. Plant Soil 234, 15-26

Trujillo M.E.; Willems, A.; Abril, A.; Planchuelo, A.M.; Rivas, R.; Ludena, D.; Mateos, P.F.; Molina, E.M. & Velazquez, E. (2005) Nodulation of *Lupinus albus* by strains of *Ochrobactrum lupini* sp. Nov. Appl. Environ. Microbiol. **71**, 1318-1327

Unkovich, M.J.; Pate, J.S. & Hamblin, J. (1994) The nitrogen economy of broadacre lupin in southwest Australia. Australian Journal of Agricultural Research, **45**, 149-164

Urquiaga, S. & Boddey, R.M. (2001) Biological nitrogen fixation associated with tropical pasture grasses. Australian Journal of Plant Physiology, **28** (9), 837-844

Urquiaga, S. & Zapata, F. (2000) Fuentes alternativas para la fertilización nitrogenada de cultivos anuales. Urquiaga, S. & Zapata, F. (eds.). Manejo eficiente de la fertilización nitrogenada de cultivos anuales en America Latina y el Caribe. Gênesis, Porto Alegre. 57-76

Vanlauwe, B.; Sanginga, N. & Merckx, R. (1998) Recovery of *Leucaena* and *Dactyladenia* residue nitrogen-15 in alley cropping system. Soil Sci. Soc. Am. J. **62**, 454-460

Virginia, R.A. & Jarrell, W.M. (1983) Soil properties in a mesquite-dominated Sonoran desert ecosystem. Soil Sci. Soc. Am. J. **47**, 138-144

Willems, A. (2006) The taxonomy of rhizobia: an overview. Plant Soil **287**, 3-14

Valverde, A.; Velazquez, E.; Santos, F.F.; Vizcaino, N.; Rivas, R.; Mateos, P.F.; Molina, E. M.; Igual, J.M. & Willems, A. (2005) *Phyllobacterium trifolii* sp. nov., nodulating *Trifolium* and *Lupinus* in Spanish soils. Int. J. Syst. Evol. Microbiol. **55**, 1985-1989

Vandamme, P.; Goris, J.; Chen, W.M.; de Vos, P. & Willems, A. (2002) *Burkholderia tuberum* sp. nov. and *Burkholderia phymatum* sp. nov. nodulate the roots of tropical legumes. Syst. Appl. Microbiol. **25**, 507-512

Potential of organic substrates in lowland rice-based systems of Asia

Mathias Becker

INRES Department of Plant Nutrition, Karlrobert-Kreiten-Strasse 13, 53115 Bonn, Germany, E-Mail: mathias.becker@uni-bonn.de

2.0 Abstract

The use of organic substrates in lowland rice-based systems of Asia was common and widespread until the mid 20th century. Despite numerous benefits, the use of organic matter applications in rice-based systems of Asia continues to decline ever since. This paper reviews the potential of various organic substrates in the diverse rice-based production systems of Asia. It concludes that the future use of organic substrates is likely to be limited to few socio-ecological niche environments, mainly those with marginal attributes or those targeting the organic market segment.

Keywords: Aluminium toxicity, *Azolla*, green manure, legumes, N_2 fixation, straw

2.1 Introduction

Concerns mount over agricultural sustainability as most Asian countries have moved into a post-Green Revolution phase (Cassman & Pingali, 1994). During the past three decades, cropping systems have evolved from a single low-input crop of rainfed lowland rice to irrigated multiple rice cropping with high external input use or intensified rainfed systems with lowland rice during the rainy season and wheat or high-value cash crops during the dry season (Becker et al., 2008; Garrity & Flinn, 1988; Ladha et al., 2003; Mussgnug et al., 2006; Prasad, 2005). A widely observed decline of partial factor productivity and in some instances even declining yield levels have been associated with intensified lowland cultivation, a poor management of external inputs, and the intensified cultivation in marginal environments (Dobermann et al., 2002; Dawe et al., 2003). When faced with stagnating or declining crop yield, farmers traditionally opted for the application of organic substrates (King, 1911). Their uses rapidly and steadily declined since the 1960's with the advent of the green revolution and the wide availability of mineral fertilizers that rapidly replaced organic substrates, even in traditional "organic" countries such as Vietnam or the Peoples Republic of China. There was a renewed interest and substantial research investments in organic sources and their use in rice following the oil crises in the mid 1970's (IRRI, 1988; Ladha & Garrity, 1994; Roger & Watanabe, 1986 Dawe et al., 2003), further boosted with the recognition of declining productivity trends in intensified production systems in the early 1990's (Cassman & Pingali, 1994; Dawe et al., 2000; Regmi et al., 2002) and more recently with trade liberalization agreements, opening new markets for organically-grown products (Haccius et al., 1995, Schmid et al., 2004). While research has substantially contributed to our understanding of processes related to the production, application and the management of organic substrates and numerous benefits for rice based systems have been reported (reviewed by Becker et al., 1995 and Becker, 2003), their use in farmers' fields of Asia continues to decline (Garrity & Becker, 1994, except for some niche environments and farmers rather opt for an ever-increasing use of external inputs (Olk et al., 2007). Have Asian farmers forgotten an important part of their (agri)cultural heritage, or are organic substrates indeed a thing of the past? This paper reviews the potential of various organic substrates in the diverse rice-based production systems of Asia and attempts to define socio-ecological niches for a future site- and system-specific targeting of organic substrate uses. The information presented is based on the available review and original papers as well as on a number of unpublished studies.

2.2 Diversity and evolution of rice-based systems

The type of organic amendments, the reasons for their application and their effect and effectiveness depend on the rice-based production system. These have evolved over time from the domestication of rice and traditional low-input subsistence production to high-input market-oriented precision agriculture in peri-urban or otherwise infrastructurally favoured environments. The key feature driving this evolution is the availability of the production factors land, labour, capital and know-how (Figure 1).

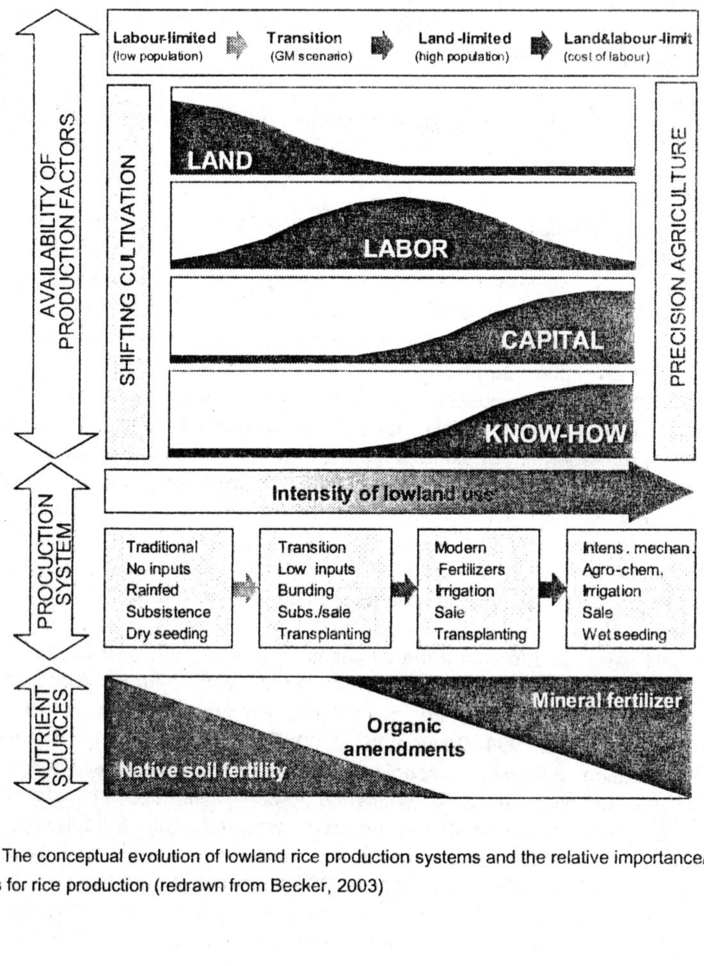

Figure 1: The conceptual evolution of lowland rice production systems and the relative importance/availability of N sources for rice production (redrawn from Becker, 2003)

In the traditional rainfed lowland systems, only a minor proportion of the available land is used. Agricultural production is limited by the availability of labour rather than land. A low population density relative to the resource base, subsistence-oriented production objectives and a reliance on household labour characterizes these traditional extensive land use systems. Only parts of the lowland area are used for growing a single crop of dry broadcast or dibble seeded long-duration rice cultivars, often in rotation with natural fallow. Land preparation is frequently done by slash-and-burn and only panicles are harvested. The crop has to rely on the native soil nutrient supply for its nutrition. This production environment characterizes large areas in Africa and is still found in few remote rural areas in Southeast Asia.

Land availability is not limiting the production. The challenge for farmers is rather how best to allocate labour between different enterprises. A substitution of labour by capital is restricted by the unavailability of cash in the subsistence-oriented production. An investment in any type of organic amendment is not attractive, as labour is limited. Increasing the area of production can easily counteract low yields resulting from unfavourable ecological conditions or from N deficiency.

During the subsequent transitional stage, the lowlands start to come under pressure from rising population levels. Lowland farmers introduce bunding as the first step towards improved water control and to maximise the output from the increasingly limiting resource land. The availability of labour permits the shift from dry seeding to transplanting of rice. Where they can, producers turn increasingly to the market. The availability of both land and labour combined with a still restricted access to factor markets for the purchase of mineral fertilizers are conditions favouring the use of organic amendments in this development stage. Today, environments with a single crop of transplanted rainfed lowland rice are restricted to remote and infrastructurally unfavoured regions such as hilly areas or smaller islands.

Beyond this stage, pressure on land for agricultural goods, but also for settlement and other non-agricultural activities continues to rise as can be observed throughout much of South and Southeast Asia. Production is once again limited by labour as emerging markets in the growing urban centres increase the cost of agricultural labour. Land and labour are increasingly substituted for by capital such as external inputs (fertilizers, pesticides) and mechanization. Direct wet seeding combined with an intensified use of herbicides replaces the transplanting of lowland rice as it has happened in most of the major rice-growing areas of the region (Mekong Delta in Vietnam, Central Luzon in the Philippines, Java in Indonesia, etc.). At the same time, the developing infrastructure improves access to agricultural extension and other means for the acquisition of know-how. Rice double- and triple-cropping systems dominate the humid tropical zone as well as the fully irrigated areas while rice-wheat and rice-vegetable rotations emerge in the partially irrigated subtropical and the rainfed tropical environments. Neither the land required for the production nor the labour required for the application of organic amendments are available. Mineral fertilizers replace organic amendments to meet the crops' nutritional requirements, pesticides substitutes for possible sanitation effects (e.g. nematode control or weed suppression) by organic substrates, and soil physical improvements may be addressed by secondary raw materials from the urban centre or by animal manures from intensified peri-urban rearing systems. The only exceptions are systems producing for the emerging markets for organically produced food. Hence, the changing availability of production factors drives not only the evolution of the rice-based production systems, it also determines the type, amount, effects and the niches for organic amendments.

2.3 Organic substrates and their effects

Various organic substrates are potentially available for use in rice-based systems. Rice straw is the sole organic amendment that is routinely applied to a significant area of lowland rice (Olk et al., 2007). Green manure legumes and the floating fern Azolla have all but disappeared and are currently still used on an estimated 1 million ha corresponding to slightly less than 1% of the lowland rice area of Asia (Becker, 2003; Roger & Watanabe, 1986). Farmyard manure, various composts and biogas sludge may be available in significant amounts in some environments but are nearly exclusively applied to higher-value upland crops grown in rotation with rice (Becker et al., 2008; Shresta & Ladha, 2000). The straw of wheat and residues of a few other upland crops are applied in some of the subtropical rice-wheat rotation systems in South and East Asia (Pande & Becker, 2003; Samra et al., 2003). The diversity of the substrates is matched by the diversity of main effects or reasons for their use.

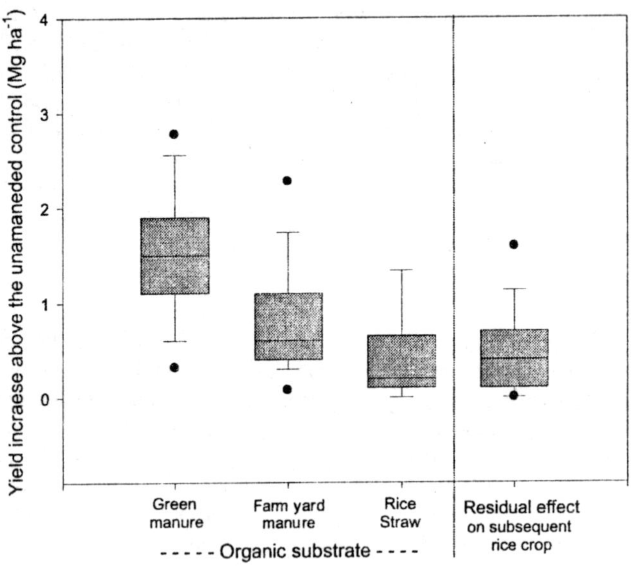

Figure 2: Yield increases by various organic substrates and their residual effects on a subsequent rice crop (redrawn from: Alberto et al. 1996, Aulakh et al., 2000; Becker 2003, Garrity & Becker, 1995; Mussgnug et al., 2006, Olk et al. 2007.

The primary desired effect of an organic matter application is to increase the grain yield of rice. Reported yield increases range from 0 to about 3 Mg ha^{-1} above an unamended control, depending on substrate type and application rates (Figure 2). Other effects may comprise the addition of nutrients, soil or fertilizer N conservation, soil physical (penetration, friability) or chemical improvements (CEC, pH, Al toxicity), or the disposal of an undesired waste product. Particularly the use of composts to upland crops can result in significant carry-over effects on the yield of the subsequent crop of lowland rice. However, the large variation of both the direct and residual effects indicates the uncertainty in the outcome of farmers' investment and the inherent riskiness of any organic substrate application strategy. The efficacy of organic substrate uses and their desired effects are highly dependent on the bio-physical environment and the prevailing production system as illustrated thereafter.

2.3.1 Substitution of mineral fertilizer N

Niches for organic substrate use in lowland rice are conditions where the availability of land and labour suffices for investment in the production and application of the organic substrate and where mineral fertilizer is either unavailable or costly or where organic substrates out-compete mineral N sources with regard to N use efficiency. At average application rates, organic substrates and mineral fertilizer N show similar agronomic N use efficiencies. However, a differentiation of 228 considered data points from the published literature into irrigated and rainfed production environments shows that urea-N tends to out-compete organic N in irrigated systems while organic N appears more efficient than urea-N in rainfed lowland rice (Figure 3).

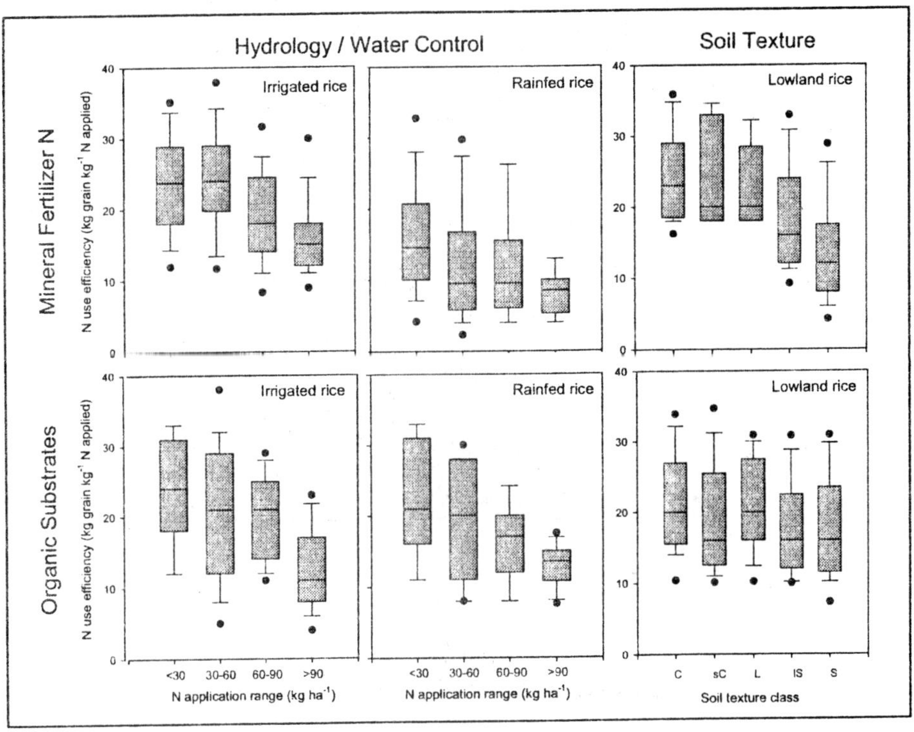

Figure 3: Comparative assessment of the use efficiency of organic and mineral N sources in lowland rice differentiated by hydrology and soil texture (redrawn from: Becker et al. 1995, Becker 2003, Dawe et al. 2003, Garrity & Becker 1995; Olk et al. 2007; Yadvinder Singh et al., 2004.

A further differentiation of the same data set indicates that N use efficiency is similar for both N sources on clay soils, but in sandy soils it tends to be substantially greater for organic substrates. Thus, an application of organic substrates can compete with mineral N fertilizers in some unfavourable sandy-textured rainfed situations as they occur in Northeast Thailand, Laos, and Kalimantan.

2.3.2 N conservation by Azolla or wheat straw

In the low-input rice-upland crop rotation systems, soil aeration status typically changes from aerobic (upland crop) to anaerobic (lowland rice) with implications on soil nitrate N dynamics (Figure 4). Thus, nitrate can build up in the soil during the dry-to-wet season transition period, which is subsequently lost upon soil flooding by leaching and denitrification (Figure 5). This system is exemplified by the rice-wheat production system of Nepal, where the N nutrition of both crops is largely based on the supply from soil pools. The return of the straw of the preceding wheat crop can result in a temporary immobilization of soil N in the microbial biomass, thus reducing the soil nitrate build-up between the harvest of wheat and the transplanting of rice with subsequent reductions in nitrate leaching and denitrification and substantial rice yield increases (Table 1).

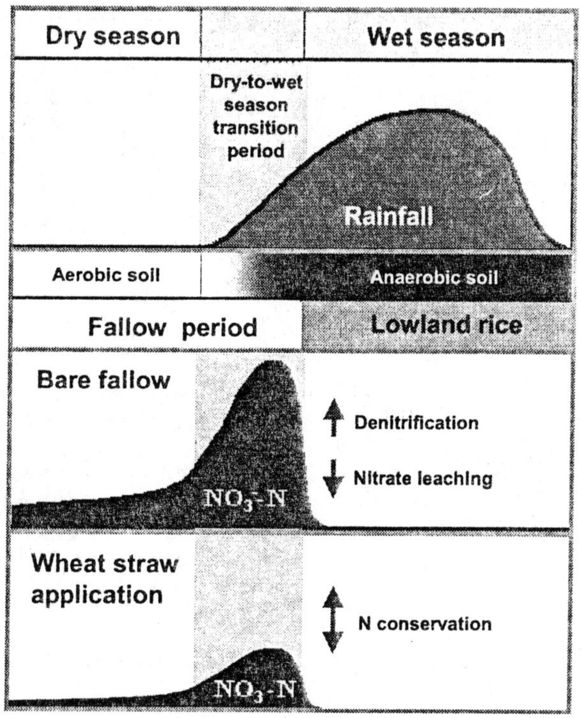

Figure 4: Conceptual presentation of soil nitrate dynamics between the dry and the wet season in lowland rice-upland crop rotations (redrawn from George et al., 1993; Pande and Becker, 2003).

While the option represented in Table 1 is restricted to the marginal high-altitude systems of the Himalayas, where a short transition period does not allow for the cultivation of high-value crops (mainly pulses and potatoes) and only little wheat straw is required as bedding material for the few animals in these subsistence-oriented small-scale farms, similar trends and effects of soil N conservation by organic residue amendments have been reported from various rice-based production systems (Phongpan & Mosier, 2003; Shresta & Ladha, 2000; Verma, 1992; Yadvinder Singh et al., 2004).

Table 1: Effect of wheat straw application during the dry-to-wet season transition period on N losses and the grain yield of rice at a high-altitude site in Nepal (redrawn from Becker et al., 2007)

Transition season treatment	Transition season			Wet season
	Soil N_{min} peak	N_2O emission	NO_3 leaching	Rice yield
	------------ (kg N ha^{-1}) ------------			($Mg\ ha^{-1}$)
Bare fallow (control)	37 a	4.2 a	6.3 a	1.47 c
Wheat straw mulch (3.0 Mg ha^{-1})	24 b	1.4 b	3.9 b	1.57 bc
Straw incorporation (3.0 Mg ha^{-1})	18b	0.9 b	0.9 c	2.53 a

Apart from nitrate N losses during the dry-to-wet transition period in seasonally flooded lowlands, ammonia volatilization is another important loss mechanism, particularly of mineral fertilizer N from irrigated rice soils. It is promoted by high floodwater concentrations of ammoniacal N, high temperatures, and an elevated floodwater pH. Recent evidence from field experiments in Nepal and the Philippines underlines the potential of *Azolla* to not only reduce N losses and increase fertilizer N use efficiency but also to significantly increase the grain yield of lowland rice, particularly in soils of relatively high pH. Dense cover of the floodwater by *Azolla* provides cooling shade and curbs the photosynthetic activity of floodwater organisms, which would dampen diurnal pH fluctuations. This muting of variations in temperature and diurnal pH, combined with ammoniacal-N interception by *Azolla*, can reduce N losses and save N in the green manure biomass for later use by rice (Vlek et al., 1995; Cisse & Vlek, 2003; De Macale, 2002; Tuladhar, 2003). The relevance of such strategies at farm level remains questionable, not least because of the large P requirements of *Azolla* and their effectiveness being limited to alkaline soils.

2.3.3 Physical and chemical improvement of soil parameters

Rice straw is the sole organic amendment that is routinely applied to significant areas of lowland rice. It is normally cut at least 20 cm above the ground, leaving large amounts of straw biomass in the field. The remaining straw is generally burned, but significant portions are also incorporated, offering significant potential benefits given its high contents of C, K and silicon (Olk et al., 2007). Yield benefits tend to be low and variable (Figure 2). Straw incorporation can result in higher grain yield than straw removal or burning, provided that incorporation was continued for a sufficient number of crop seasons (Ponnamperuma, 1984). The benefit is greater in warmer climates, where toxic compounds released by incorporated straw will decompose quickly before transplanting of the next crop (Cho & Ponnamperuma, 1971). Particularly in marginal soils, such as the strongly degraded soils on the upper terraces of the large Asian rivers (i.e. Red River Delta), straw application can temporarily increase the CEC, which in combination with substantial K addition by straw resulted in significant yield increases (Mussgnug et al., 2006). Soil physical properties can also be improved, including water percolation and ease of puddling, porosity, water retention, and bulk density (Walia et al., 1995). Such effects tend to be short-lived due to the low lignin content of 5-10% (Becker et al., 1994) and the resulting rapid mineralization of rice straw. Straw incorporation has not always proven advantageous. Thus straw did not improve grain yield trends in seven long-term trials of continuous rice and 18 trials of rice-wheat in Asia (Dawe et al., 2003). In addition, straw incorporation has been shown to result in substantial methane emissions (Alberto et al., 1996). Finally, incorporation requires machinery that is often not available and can result in the accumulation of toxic substances (Tsutsuki, 1984), and the temporary immobilization of N (Shibahara et al., 1998). In practice, economic constraints often promote straw burning or removal from the field.

Another type of widely available organic substrates is derived from decentralized waste/water management. These comprise anaerobic sludges (i.e. from biogas plants) as well as different aerobic composts. Such substrates are of particular interest in parts of China and Vietnam. A low productivity of rice-based systems in the Mekong Delta is linked to low N use efficiency and low soil organic matter content in alluvial soils and to Al toxicity and P deficiency in acid sulfate soils. The increasing diversification of the cropping system by replacing the dry season rice by high-value horticultural crops is likely to further exacerbate these soil-related problems. The organic substrates from waste and wastewater treatment are increasingly used in these systems to address the widely observed declining productivity trends. Several field studies could show that biogas sludge application increased the dry matter accumulation of diverse upland crops by 25-98% above the unfertilized control. The effect was associated with an improved N supply grown on a degraded alluvial soil (Figure 5, left).

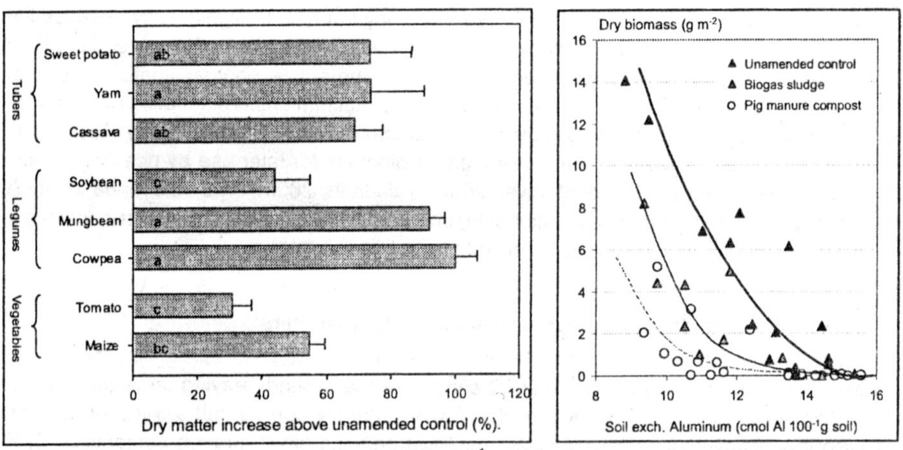

Figure 5: Response to biogas sludge application (6 Mg ha^{-1}) of upland crops grown in rotation with lowland rice on a degraded alluvial soil of the Mekong Delta (left) and organic substrate effects on biomass accumulation and Al^{3+} tolerance of *Vigna radiata* L. grown on acid sulphate soils in rotation with lowland rice (right) (source: Becker et al., 2008).

A favourable upland crop response to an application of P-rich aerobic compost in an acid sulphate soil was mainly related to an alleviation of H^+ and Al^{3+} toxicities (Figure 5, right). Such effects were much less with anaerobic substrates such as biogas sludge. Neither sludge nor composts produced significant yield effects under favourable (Molisol) soil conditions (Becker et al., 2008).

2.4 Conclusions

While being highly desirable, the effectiveness of organic amendment in rice-based systems is determined by both biophysical and socioeconomic factors. The use of organic substrates is generally limited by land availability for the production of the organic biomass (not applicable for straw), and by labour availability for the application and incorporation into the rice field. From the examples presented here, the effectiveness of different organic amendments may have additional restrictions to specific niche environment (Table 2).

Table 2: Organic amendments, their primary effects and the socio-ecological niches for their use

Source of organic amendment	Primary effect	Biophysical niche	Socio-economic niche
Green manure legumes	BNF-N addition	Rainfed rice Light-textured soils	Availability of land and labour
Azolla	Reduced NH_3 volatilization	Irrigated rice Alcaline soils, high P	Availability of labour
Rice straw	K addition	Upland crop rotations Degraded soils	Availability of labour / implements
Wheat straw	Soil N conservation	Rice-wheat rotations High-altitude systems	No competing uses of straw
Compost	Nutrient addition Buffer Al toxicity	High-value crops Acid sulphate soils	Availability of labour
Biogas sludge	CNK addition	Upland crop rotations Degraded / alluvial soils	Availability of labour / implements
Farmyard manure	Carbon / nutrient addition	All environments	Availability of labour

Thus green manure legumes show highest competitive performance compared to mineral fertilizers in rainfed environments with sandy textured soils. Composts, sludges and manure are particularly interesting to correct element imbalances and toxicities in problem soils. While being an excellent organic substrate for a wide range of environmental conditions, farmyard manure is often not available in sufficient amounts, as draft animals are no longer widely used. Its use in rice is often not economical because of high labour and transport costs.

As a result, animal manure is usually applied to high-value crops. Benefits of rice straw application are mainly limited to K cycling in degraded soils and the use of wheat straw in rice-wheat rotations can effectively contribute to soil N conservation only in marginal high altitude systems. Finally, Mineral N savings and N additions by Azolla are restricted to alkaline soils and conditions of high P availability, respectively.

It may be concluded that the future use of organic substrates in rice-based systems is likely to be limited to few socio-ecological niche environments, mainly those with marginal attributes or those targeting the organic market segment, where higher commodity prices justify additional land, labour and capital investments.

References

Alberto, M.C.R.; Neue, H.U.; Capati, A.B.; Castro, R.U.; Bernardo, L.M.; Aduna,J.B. & Lantin R.S. (1996) Effects of different straw management practices on soil fertility, rice yields, and the environment. Attanandana, T.; Kheoruenromne, I.; Pongsakul, P. & Vearasilp, T. (eds.) Maximizing sustainable rice yields through improved soil and environmental management. Thailand Department of Agriculture, Bangkok, Thailand, 721-731

Aulakha, M.S.; Khera, T.S.; Doran, J.W. & Bronson, K.F. (2000) Managing crop residue with green manure, urea and tillage in a rice–wheat rotation. Soil Sci. Soc. Am. J. **65**, 820-827

Becker, M.; Ladha, J.K.; Simpson, I.C. & Ottow, J.C.G. (1994) Parameters affecting residue nitrogen mineralization in flooded soils. Soil Sci. Soc. Am. J. **58**, 1666-1671

Becker, M. (2003) Potential von Leguminosen zur Gründüngung und Einsatzmöglichkeiten im tropischen Reisanbau. BOD Publishers, Norderstedt, Germany

Becker, M.; Asch, F.; Chiem, N.H.; Ni, D.V.; Saleh, E.; Tanh, K.V. & Tinh, T.K. (2008) Decomposition of organic substrates and their effect on mungbean growth in two soils of the Mekong Delta. J. Agric. Rural Develop. Trop. Subtrop. **109**(2) in press

Becker, M.; Ladha; J.K. & Ali, M. (1995) Green manure technology: Potential usage, limitations. A case study for lowland rice. Plant and Soil. **174**,181-194.

Becker, M.; Maskey, S.L.; Pande, K.R.; Shah, S.C. & Shrestha, S.C. (2007) Effects of transition season management on soil N dynamics and system N balances in rice-wheat rotations of Nepal. Field Crops Res. **103**, 98-108

Cassman, K.G. & Pingali, P.L. (1994) Extrapolating trends from long-term experiments to farmers' fields: The case of irrigated nice systems in Asia. Varnett, P.; Payne, V. & Steiner, R. (eds.) Agricultural Sustainability: Economic, Environmental and Statistical Considerations. London: John Wiley and Sons, Ltd., London, UK, 63-86

Cho, D.Y. & Ponnamperuma, F. (1971) Influence of soil temperature on the chemical kinetics of flooded soils and the growth of rice. Soil Science **112**, 184-194

Cisse, M. & Vlek, P.L.G. (2003) Conservation of urea-N by immobilization remobilization in a rice-Azolla intercrop. Plant Soil **250**, 105-112

Dawe, D.; Dobermann, A.; Moya, P.; Abdulrachman, S.; Singh, B.; Lal, P.; Li, S.Y.; Lin, B.; Panaullah, G.; Sariam, O.; Singh, Y.; Swarup, A.; Tan, P.S. & Zhen, Q.X. (2000) How widespread are yield declines in long-term rice experiments in Asia? Field Crops Res. **66**, 175-193

Dawe, D.; Dobermann, A.; Ladha, J.K.; Yadav, R.L.; Bao, L.; Gupta, R.K.; Lal, P.; Panaullah, G.; Sariam, O.; Singh, Y.; Swarup, A. & Zhen, Q.X. (2003) Do organic amendments improve yield trends and profitability in intensive rice systems? Field Crops Research. **83**, 191-213

Dobermann, A.; Dawe, D.; Reimund P.; Roetter, R.P. & Cassman, K.G. (2000): Reversal of Rice Yield Decline in a Long-Term Continuous Cropping Experiment. Agron. J. **92**, 633-643

De Macale, M.A.R. (2002) The role of Azolla cover in improving the nitrogen use efficiency of lowland rice. Center for Development Reserach (ZEF), Ecology and Development Series, No 2, Cuvillier Verlag, Göttingen, Germany

Garrity, D.P. & Becker, M. (1995) Where do green manures fit in Asian rice farming Systems? Ladha, J.K. & Garrity, D.P. (eds.) Green manure production systems for Asian ricelands. The International Rice Research Institute, Manila, Philippines, 2-12

Garrity, D.P. & Flinn, J.C. (1988) Farm-level management systems for green manure crops in Asian rice environments. Sustainable Agriculture: Green Manure in Rice Farming. The International Rice Research Institute, Los Banos, IRRI, Philippines, 111-129

George, T.; Ladha, J.K.; Buresh, R.J. & Garrity, D.P. (1993) Nitrate dynamics during the aerobic soil phase in lowland rice-based cropping systems. Soil Sci. Soc. Am. J. **57**, 1526-1532

George, T.; Ladha, J.K.; Garrity, D.P. & Torres, R.O. (1995) Nitrogen dynamics as nitrate catch crops during dry to wet transition in lowland rice cropping systems. Agron. J. **86**, 267-273

Haccius, M.; Bernd, A. & Geier, B. (1995) International IFOAM Conference on Trade in Organic Products, 4. International Federation of Organic Agriculture Movements, Tholey-Theley, Frankfurt, Germany, 131 pp.

IRRI – International Rice Research Institute (1988) Sustainable Agriculture: Green manure in rice farming. The International Rice Research Institute, Manila, Philippines

King, F.H. (1911) Farmers of Forty Centuries; or, Permanent Agriculture in China. Korea and Japan. Emmaus, PA Rodale Press. London, U.K

Ladha, J.K.; Dawe, D.; Pathak, H.; Padre, A.T.; Yadav, R.L.; Singh, B.; Singh, Y.; Singh, Y.; Singh, P.; Kundu, A.L.; Sakal, R.; Ram, N.; Regmi, A.P.; Gami, S.K.; Bhandari, A.L.; Amin, R; Yadav, C.R.; Bhattarai, E.M.; Das, S.; Aggarwal, H.P.; Gupta, R.K. & Hobbs, P.R. (2003) How extensive are yield declines in long-term rice-wheat experiments in Asia? Field Crops Res. **81**, 159-180

Ladha, J.K. & Garrity, D.P. (1994) Green manure production systems for Asian ricelands. International Rice Research Institute, Manila, Philippines

Mussgnug, F.; Becker, M.; Son, T.T; Buresh, R.J. & Vlek P.L.G. (2006) Yield gaps and nutrient balances in intensive, rice-based cropping systems on degraded soils in the Red River Delta of Vietnam. Field Crops Res. **98**, 127-140

Olk, D.; Becker, M.; Linquist, B.A.; Pandey, S. & Witt, C. (2007) Integrated nutrient management - Experiences/concepts from Southeast Asia. Aulakh M., Olk D. (eds.) Integrated nutrient management for sustainable agriculture. Haworth Publishers, Brandon, Canada

Pande, K.R. & Becker, M. (2003) Seasonal soil nitrogen dynamics in rice-wheat cropping systems of Nepal. J. Plant Nutr. Soil Sci. **166**, 499-506

Phongpan, S. & Mosier, A.R. (2003) Effect of crop residue management on nitrogen dynamics and balance in a lowland rice cropping system. Nutr. Cycl. Agroecosyst. **66**, 133-142

Ponnamperuma, F. (1984) Straw as source of nutrients for wetland rice. IRRI (ed.) Organic matter and rice. International Rice Research Institute Manila, Philippines, 117-136

Prasad, R. (2005) Rice-wheat cropping systems. Adv. Agron. **86**, 255-339

Regmi, A.P.; Ladha, J.K.; Pathak, H.; Pashuquin, H.E.; Bueno, C.; Dawe, D.; Hobbs, P.R.; Joshy, D.; Maskey, S.L. & Pandey, S.P. (2002) Yield and soil fertility trends in a 20-year rice-rice-wheat experiment in Nepal. Soil Sci. Soc. Am. J. **66**, 857-867

Roger, P.A. & Watanabe, I. (1986) Technologies for utilizing biological nitrogen fixation in wetland rice: Potentialities, current usage and limiting factors. Fert. Res. **9**, 39-77

Samra, J.S.; Singh, B. & Kumar, K. (2003) Managing crop residues in the rice-wheat system of Indo Gangetic Plain. Ladha, J.K.; Hill, J.E.; Duxbury, J.M.; Gupta, R.K. & Buresh, R.J. (eds.) Improving the productivity and sustainability of rice-wheat systems, issues and impacts, ASA Spec. Publ. No. 65, 173-195

Schmid, O.; Sanders, J. & Midmore, P. (2004) Organic Marketing Initiatives and Rural Development. OMIaRD Publication Volume 7. University of Wales, Aberystwyth, Australia.

Shibahara, F.; Yamamuro, S. & Inubushi, K. (1998) Dynamics of microbial biomass nitrogen as influenced by organic matter application in paddy fields. I. Fate of fertilizer and soil organic N determined by ^{15}N tracer technique. Soil Science and Plant Nutrition **44**, 167-178

Shrestha, R.K. & Ladha, J.K. (2000) Recycling of residual soil nitrogen in a lowland rice-sweet pepper cropping system. Soil Sci. Soc. Am. J. **64**(5), 1689-1698

Tsutsuki, K. (1984) Volatile products and low-molecular-weight phenolic products of the anaerobic decomposition of organic matter. IRRI (ed.) Organic matter and Rice. International Rice Research Institute, Manila, Philippines, 329-343

Tuladhar, K.J. (2003) The effect of Azolla on nitrogen use efficiency in rice-wheat rotatioons of Nepal. Center for Development Reserach (ZEF), Ecology and Development Series, No 13, Cuvillier Verlag, Göttingen, Germany

Verma, T.S. & Bhagat, R.M. (1992) Impact of rice straw management practices on yield, nitrogen uptake and soil properties in a wheat rice rotation in northern India. Fert. Res. **33**, 97-106

Vlek, P.L.G.; Diakite, M.Y. & Mueller, H. (1995) The role of Azolla in curbing ammonia volatilization from flooded rice systems. Fertil. Res. **42**, 165-174

Walia, S.S.; Brar, S.S. & Kler, D.S. (1995) Effects of management of crop residues on soil properties in rice-wheat cropping system. Environ. Ecol. **13**, 503-507

Yadvinder-Sing; Bijay-Singh; Ladha, J.K.; Khind, C.S.; Gupta, R.K.; Meelu, O.P. & Pasuquin, E. (2004) Long-term effects, of organic inputs on yield and soil fertility in the rice-wheat rotation. Soil Sci. Soc. Am J. **68**, 845-853

Organic vegetable production and research needs in the tropics and subtropics

[1]Peter Juroszek, [2]Thomas A. Lumpkin, [3]Warwick Easdown and [1]Hsing-Hua Tsai*

AVRDC-The World Vegetable Center, Yi-Ming Liao 60, Shanhua 74151, Taiwan
[1]Crop & Ecosystem Management Unit, Organic Vegetable Program
[2]Director General
[3]Head of Communication and Training Office
**Corresponding author: juroszek@avrdc.org*

3.0 Summary

Fruits and vegetables are among the major sources of micronutrients for the majority of the world's poor living in the tropics and subtropics. While improving organic vegetable production in these regions can help the small number of certified organic growers who produce mainly for high-priced niche export markets, such an effort would also have wider implications. In sub-Saharan Africa the majority of farmers are 'organic by default' because they cannot afford chemical inputs; therefore, many of the systems developed for certified organic farmers also may be applicable to the greater farming community.

Tropical farmers producing vegetables organically, whether by design or default, must overcome significant challenges organic growers in temperate climates seldom face, including a lack of suitable varieties, heavy rainfall, the year-round presence of pests and pathogens, and abiotic stressors such as heat, drought, tropical cyclones, and temporary flooding. Successful vegetable production requires a minimum level of inputs, and in many developing countries in the tropics and subtropics organic input markets are poorly developed. However, local markets for organic products are growing, fuelled by rising incomes, urbanization, and increasing consumer concern about the excessive and unregulated use of pesticides by conventional vegetable farmers, particularly in Asia.

There has been a severe lack of accessible research papers on organic vegetable production in the tropics and subtropics. Only a few organically-grown vegetable crops have been researched worldwide, including tomato, lettuce, carrot and cucumber; most of this research has been done in developed countries under temperate and subtropical climatic conditions. Vegetable crops important in the tropics and subtropics such as eggplant have not yet been addressed, nor have indigenous vegetables. Because of their adaptation to tropical and subtropical environments, and often good tolerance of pests, diseases and low soil fertility, some indigenous vegetables may be more suited to regional organic vegetable production systems than introduced 'exotic' vegetables. Institutionalized research in organic vegetable production in most tropical and subtropical countries is relatively new, and with the exception of AVRDC – The Word Vegetable Center, is not a significant focus for the International Agriculture Research Centers. More institutionalized research in tropical and subtropical Organic Agriculture and horticulture may need to be promoted, coordinated, and funded on an international level similar to a CGIAR Challenge Program.

There is a major need for improved systems to manage pests, pathogens and low soil fertility in tropical and subtropical organic vegetable production. Only limited work has been done on the use of botanical pesticides despite the traditional use of mixtures containing plants such as neem for pest control in the tropics and subtropics. Organic vegetable farmers often lack access to superior varieties or quality seed. Plant breeding that produces adapted high-yielding varieties with good disease and pest tolerance is particularly needed. Much of the farmland in the tropics and subtropics, particularly in Africa, is degraded or has low fertility. Plant breeding traditionally has focused on high-fertility conditions, but it also will need to take into account low-fertility soils, particularly those that are P-deficient.

With appropriate research to develop profitable and adapted organic vegetable production in the tropics and subtropics, the industry could expand beyond its current niche role and may contribute to a lower environmental impact. The growing local demand for organic vegetables could be met, and conventional farming systems that may be "organic by default" could also benefit.

Keywords: soil fertility, crop nutrition, pest management, natural resources

3.1 Introduction

Definition of Organic Agriculture

Organic Agriculture is no longer just a phenomenon of developed countries. It is now commercially practiced in 120 countries, representing 31 million ha of certified cropland and pasture with a market of US$40 billion in 2006 (Willer & Youssefi, 2007). The Food and Agriculture Organization of the United Nations (2005) defines Organic Agriculture as:

> *...a holistic production management system which promotes and enhances agro-ecosystem health, including biodiversity, biological cycles and soil biological activity. It emphasizes the use of management practices to achieve this in preference to the use of off-farm inputs, taking into account that regional conditions require locally adapted systems. This is accomplished by using, where possible, agronomic, biological, and mechanical methods, as opposed to using synthetic materials, to fulfill any specific function within the system.*

In general, any genetically modified organism (GMO) is "not welcome" in Organic Farming systems (IFOAM, 2004).

In practice there is not a clear-cut distinction between conventional and Organic Farming, as the term "conventional" masks a great diversity of commonly used farming management practices (Kristiansen, 2006), many of which are also used by organic farmers. For instance, a conventional grower may use both mineral fertilizers and green manures while avoiding synthetic pesticide applications.

Organic Farming is not necessarily a low-input farming system – particularly as it applies to vegetables. Both low-input and high-input systems can be found in organic and conventional farming systems. Vegetables generally need relatively high inputs of nutrients and plant protection interventions to reach satisfactory levels of yield and quality. An increasing number of organic farmers are sourcing their organic inputs off-farm. There is now a growing input market for organic fruit and vegetable production, particularly in countries such as the USA (Caldwell et al., 2005). Many organic vegetable farming systems worldwide may be considered to be high-input systems.

For some farmers the cost of mineral fertilizers or pesticides is beyond reach. These "organic-by-default" systems are very common among smallholder farmers, particularly in sub-Saharan Africa. Our focus, however, will be primarily on farmers who have a choice, and may be officially certified as organic.

Definition of (sub) tropical areas

Tropical and subtropical vegetable production can be found in lowland and upland areas. Large vegetable production areas are usually found in zones within countries as determined by market access, suitable soil types, water supplies and infrastructures as well as agroclimatic factors. In general, it is difficult to relate our statements specifically to either subtropical or tropical countries. Countries such as India include areas with subtropical and tropical climatic conditions. Therefore, within this paper we will use the term "(sub) tropics" as defined by Freyer (2007). Most of our examples will be from Africa and Asia.

Many countries within the (sub) tropics can be classed as "developing" or "emerging" but this term can hide large regional variations in the level of development. These countries can be significant vegetable producers. China is generally classified as an emerging country and has extensive areas of highly intensive vegetable production as well as many isolated and underdeveloped regions (Zhang et al., 2007). It is by far the world's largest producer of fruits and vegetables, accounting for about a third of total world production (FAO, 2007a).

Organic vegetable farming in the (sub) tropics faces unique challenges compared to organic production in temperate areas. These include higher pest and disease pressure that is often present year-round, a lack of adapted varieties, limited markets for organic inputs, a predominance of smallholder production often with limited access to technical and financial resources, and a reliance on export rather than domestic markets for produce sales. Organic vegetable production tends to be limited to a small number of crops for which there is an assured market, such as tomato, pepper, cabbage, onion, beans, and cucumber (UN, 2003). There has been little use of indigenous vegetables, which offer good domestic marketing opportunities and are often well adapted to (sub) tropical areas.

In this paper we first point out the lack of statistical data on organic vegetable production and consumption (section 2). We discuss the relative importance of organic vegetable production in (sub) tropical countries (section 3), major constraints and what is needed to develop organic vegetable production systems in these regions (section 4). We discuss the productive potential of organic systems (section 5) and then we present findings of our online literature survey related to research in organic vegetables worldwide (section 6), before we provide an overview of research needs particularly related to pest, disease and fertility management in (sub) tropical countries (section 7). We then briefly discuss the implications of the depletion of natural phosphorus resources and appropriate management strategies including breeding (section 8) before providing summary conclusions in section 9.

3.2 Lack of data on organic vegetable production and consumption

A lack of universal definitions of what constitutes Organic Farming hinders the collection of accurate production statistics. This is particularly true in (sub) tropical areas where "certified" organic farms may be in the minority compared to those that are "organic by default". Hamm (2005) pointed out that there is hardly a country in the world in which up-to-date and valid statistical data exist beyond a simple enumeration of the number of organic farms and their organically managed area. Only in a few countries in which the organic market is well developed, such as in Denmark, are detailed market data available from public and private institutions. In most developing countries statistics on Organic Farming are usually sporadic and often aggregated, so it is difficult to quantify the area under Organic Farming (Bett & Freyer, 2007) or the amount of organic vegetables produced (Saxena, 2007). Most organic market studies have analyzed organic produce in general and not the organic vegetable market per se. In sub-Saharan Africa fresh certified organic vegetables, geared mostly for export markets, are produced in Kenya, Uganda, Madagascar, Malawi, South Africa, and Zambia (Parrott et al., 2006 as cited by Saxena, 2007). Because of their higher economic value compared to other crops such as rice, vegetables are particularly important to Organic Farming in (sub) tropical areas.

3.3 The relative importance of vegetables in Organic Farming in (sub) tropical countries

Organic vegetable production is a profitable niche industry

Organic vegetable production in the (sub) tropics is currently a niche industry mainly aimed at the export market. In many (sub) tropical countries, national markets for organic vegetables have developed very slowly (UN, 2003; Saxena, 2007), and most organic vegetable production is exported to high-priced markets in North America, Japan, and some European countries such as Germany and Great Britain (UN, 2003). The USA is the largest global market for organic produce with retail sales of about USD14 billion in 2005 followed by Germany with about USD5 billion in 2006. In both countries organic fruits and vegetables accounted for about 30% of total organic produce sold (Organic Trade Association, 2006 as cited in Winter & Davis, 2006; Saxena, 2007). However, in countries such as Argentina, Brazil, China, and India consumers are increasingly interested in safe food and local retailers are starting to sell organic fruits and vegetables (UN, 2003). According to Zhou (2003) vegetables are the most profitable part of the organic food market in China. The domestic market now takes 20% of China's total organic vegetable production. Prices to farmers for organic vegetables vary enormously, but range from 10 to 100% higher than those for conventional vegetables (UN, 2003; FAO 2007b).

Vegetables are a mainstay of Organic Farming activities

In Asia, vegetables are an important part of diets. Farm sizes are usually small and labour is readily available. Under these conditions vegetables can provide an important source of farm income. For example, according to Pai (2006), there were 914 certified organic farms in Taiwan in 2006. The area of organic rice was largest with 746.4 hectares followed by organic vegetables with 372.7 hectares. However, in terms of production value, organic vegetables (NTD249 million or USD7.55 million) were more important than organic rice (NTD158 million or USD4.78 million), or any other organically produced crop in Taiwan. This relationship is also true for other countries in the (sub) tropics such as Thailand (Danuwat, 2006) and

Kenya (Bett & Freyer, 2007). Because of the high labour requirement, vegetable production can benefit local economies by providing additional sources of employment (IFAD, 2005; Zhang et al., 2007); this applies to organic vegetable production in particular, as it tends to be more labour- and knowledge-intensive than conventional farming (Geisen, 1999).

A growing market for vegetables with lower pesticide residues

Pesticide poisoning is a serious global problem. Each year there are an estimated 3 million cases of acute severe pesticide poisoning reported worldwide. These include suicide attempts and accidents; the number of unreported and less severe cases of poisoning may be even higher (WHO and UNEP, 1990 as cited in Koh & Jeyaratnam, 1996).

Pesticide residues in fruits and vegetables have a large psychological impact on consumers because such foods are often eaten fresh and unprocessed. These concerns have raised the profile of organic fruits and vegetables compared to organic cereals (UN, 2003). Fruits and vegetables together account for a large share of the global pesticide market. In most years, they account for between 26-28% of total pesticide use, worth around $8.4 billion USD per year (Dinham, 2003); they tend to be overused and abused where they are affordable (Rajotte et al., 2005). Conventionally-produced fruits and vegetables more often exceed the Maximum Residue Limits (MRLs) for various synthetic pesticides (Köpke, 2003) and among vegetables the frequency of MRLs being exceeded also varies between crops (Gonzalez-Rodriquez, 2007). In Europe (EU, 2005), MLRs were exceeded most often in peppers (6%), followed by grapes (5%), cucumbers (3%) and eggplant (3%). Pesticide residues are often attributed to the failure of farmers to restrain application before harvest, the use of prohibited pesticides, and a lack of knowledge and insufficient training (Freidberg, 2003; Borght, 2004; Roitner-Schobesberger et al., 2007).

The presence of pesticide residues has significant trade implications (Weinberger & Lumpkin, 2005). Export standards introduced by the food industry, like GlobalG.A.P. (formerly EurepGAP), already present a challenge for Kenyan export-oriented horticulture, which is targeting the European market (Mausch et al., 2006). In the future, such standards may become even more important. According to review papers by Woese et al. (1997), Gennaro and Quaglia (2003), Köpke (2003), Pimentel et al., (2005), Winter and Davis (2006) and Kristiansen (2006) organic vegetables frequently contain lower nitrate levels and pesticide residues than conventionally-produced vegetables. Organic Agriculture cannot ensure that products are always completely free of synthetic pesticide residues due to general environmental pollution (FAO/WHO, 2005). Contamination is possible from highly persistent pesticides in soil and water, or contamination via pesticide-drift from neighbouring conventionally-managed fields, although drift can be minimized by planting hedgerows (Gonzalez et al., 2005). In spite of these limitations Organic Farming can help farmers meet increasingly stringent vegetable production standards.

3.4 Vegetable production constraints in (sub) tropical countries

Vegetable farmers in (sub) tropical countries are confronted with a range of production constraints including climatic stressors such as heat, seasonal drought, tropical cyclones, temporary floods and the year-round prevalence of insect pests and diseases (UN, 2003).

Both conventional and organic vegetable farmers also face other social, economic and cultural issues which constrain the growth of their industry. These were recently identified by the Global Horticulture Assessment (2005) in which more than 750 participants from 60 countries identified eight primary international issues that affect the growth of horticultural development in (sub) tropical developing countries including market systems, post-harvest systems & food safety, genetic resources conservation & development, sustainable production systems & natural resources management, capacity building, enabling environment, gender equity, and nutrition & human health.

At the farm level, Freyer (2007) identified five driving forces that lead to unsustainable land use in traditional "Organic Farming by default" systems. Constraints included: (1) a lack of basic agricultural production techniques such as water harvesting systems, (2) high labour requirements for daily survival such as collecting firewood and water, which otherwise could be allocated to agricultural production, (3) a lack of farmers' education, knowledge, and information, (4) a lack of financial resources for basic investments to sustainably manage the land, (5) and small-scale production systems that often "force" farmers to manage their agricultural land unsustainably.

At the international, national, and community levels, Kilcher (2007) identified additional constraints resource-poor farmers face on the road to Organic Agriculture. These include (6) increasing complexity of certification processes, (7) limited participation in global trade standards development, (8) limited access to national and international product markets, (9) the sometimes unfair share of income from the organic products trade, and (10) the inappropriate degree of internal organization among individual farmers. Solutions to overcome these constraints include greater investment in practice-oriented research, capacity building and extension; accessible, locally-adopted certification schemes and harmonized standards; organic market initiatives; and fair trade and good partnerships.

3.5 The productive potential of organic vegetables

Access to specific inputs and markets can be just as important as environmental and management factors in determining the yield potential of organic vegetables. Certified Organic Agriculture, which requires well-established input and output markets, may be most appropriate in regions with high production potential and access to national and international markets, such as those found in urban and peri-urban areas in (sub) tropical developing countries (Grenz & Sauerborn, 2007).

Realizing high yields is a combination of both the inherent productivity of the farm environment and the farmers' ability and willingness to achieve the yield potential on their own farms (Barker, 1979). Under highly favourable crop growing conditions, organically managed crops generally do not achieve the same yields as conventionally grown crops. For instance in conventional farming systems under optimal growing and high-input conditions about 10 t ha^{-1} grain yield can be realized in winter wheat (e.g. Sieling et al., 2005) and rice (e.g. Peng et al., 1999), and yields of up to 13 t ha^{-1} are possible (Barker, 1979). To our knowledge, this extremely high yield level cannot be realized in organic cereal production, which lacks the equivalents of readily available mineral fertilizers, effective herbicides and fungicides, and synthetic plant growth regulators commonly used to realize such extremely high yield levels. According to Tamm et al. (2007), the yield of wheat produced under Organic Farming conditions was repeatedly shown to be between 20 and 40% lower than yields achieved in comparable conventional farming systems. This is thought to be at least partly due to lower N supply and availability to the crop later in the growing season and wheat cultivars that are poorly adapted to current Organic Farming conditions.

Even with improved varieties, organic farmers' yields are normally lower than those that can be achieved under carefully controlled experimental conditions (Barker, 1979). The individual farmer's crop management skills can have a large impact even in highly productive environments. This was recently shown in tomato trials in (sub) tropical Taiwan reported by Juroszek et al. (2008a). Although there was considerable variation among all farmers, three conventional farmers realized higher and more consistent total tomato fruit yields (67.0-97.3 t ha^{-1} FW) across two processing varieties compared to their three organic counterparts (36.3-70.5 t ha^{-1} FW) where the effects of individual management were more pronounced. Careful crop management may be even more important to achieving high yields under organic compared to conventional farming conditions (Aini et al., 2005).

By improving the soil condition, Organic Farming may also benefit yield potentials. Most of the world's agricultural land area has only a medium to low yield potential because of adverse natural conditions, soil degradation, and nutrient mining (Grenz & Sauerborn, 2007). Under these conditions the productivity of Organic Farming systems can be comparable to conventional systems such as reported by Brandt and Kidmose (2002), IFAD, (2005), Ramesh et al. (2005), Kaffka et al. (2005), and Badgley et al. (2007). A key consideration is careful management of soil fertility and structure through interventions such as legume and catch crop cultivation, and compost and manure applications. As a result, Ramesh et al. (2005) state that Organic Farming may increase yields in India compared to "low-input" conventional agriculture, particularly under rain-fed production, as improved soil characteristics contribute to better water- and nutrient-holding capacity (Mäder et al., 2002).

However, most vegetables do require relative high levels of specific inputs to achieve economic yields and quality that is marketable (George & Eghbal, 2003). Therefore, severe low-input systems may not work for most vegetables. In addition, Organic Farming on a commercial scale relies on improved superior vegetable varieties with disease resistance to compensate for the lack of effective fungicides to control some economically important plant diseases. Unfortunately, in many rural areas of developing countries seeds of improved varieties often are not available at all or are in short supply (Weinberger & Lumpkin, 2005; Nath, 2007). It is important that markets to supply essential inputs are developed side-by-side with organic and conventional vegetable production systems (Rajotte et al., 2005).

Sustainability over the long term is just as important as productivity in the short term. Principles of 'Good Agricultural Practice' should be implemented in Organic Farming (e.g. Aini et al., 2005; Palada et al., 2006; Juroszek et al., 2008b). According to Kaffka et al. (2005), many people assume that Organic Farming systems are sustainable, but there is little data available on the long-term performance of organic systems. Raupp et al. (2006) summarized worldwide results and insights provided by different long-term trials in Organic Farming over several decades. Information is particularly lacking from practical organic farms or experimental systems in the (sub) tropics (Zundel et al., 2007). There are comprehensive reports available such as 'Organic Agriculture and Poverty Reduction in Asia - China and India Focus' (IFAD, 2005) and the FAO report 'Organic Agriculture and Food Security' (Scialabba EL-Hage, 2007), and these indicate that Organic Farming can be a viable approach for resource-poor farmers. However, Organic Farming systems can be very different and therefore the generalization of results is extremely difficult. In general, each system has to be adapted to the local farming, social, geographical and climatic factors to be sustainable (Kristiansen, 2006). This may mean that inspectors of international certification agencies need extensive local knowledge of the agricultural and environmental situations in (sub) tropical countries. In addition, national certification bodies in (sub) tropical countries should be established with staff members who have extensive knowledge of local agriculture and its constraints (Abele et al., 2007; Luedeling & Wichern, 2007). It may also mean that standards for certified Organic Farming should address very specific local environmental problems, such as soil salinity (Luedeling & Wichern, 2007) and soil nutrient depletion (Grenz & Sauerborn, 2007).

3.6 Research needs in organic vegetable production worldwide

The focus and extent of research

There is a severe lack of research literature on organic vegetable production, and the little that is available tends to be focused on temperate countries. The lack of publicly accessible research papers in refereed journals and conference proceedings on organic vegetable production in the (sub) tropics might be one major impediment to progress in the industry.

Research work is important to assist certified and non-certified organic farmers as well as traditional farmers who are considered to be "organic by default" because they cannot afford chemical inputs. Research should address needs that organic farmers cannot solve by themselves such as the breeding of superior vegetable varieties and the development of effective crop nutrition and pest management methods. Because successful Organic Farming is knowledge-intensive and has to be adapted to local conditions, participatory and integrated research approaches that involve farmers and other stakeholders are likely to be more appropriate (Rajotte et al., 2005; Borlaug, 2007).

Many publications are available in which the knowledge and experiences of commercial organic vegetable producers is summarized (USDA-ARS & NAL, 2003). Books on sustainable and organic vegetable production for different environments are available, including temperate (e.g. Coleman, 1995; George & Eghbal, 2003; Davies & Lennartsson, 2005), subtropical (e.g. Peet, 1996; Garrett & Beck, 1999) and tropical production conditions (e.g. UN, 2003; Aini et al., 2005). In contrast, there are fewer research papers in refereed journals and conference proceedings available on the effects of using Organic Farming methods for vegetable production, particularly related to (sub) tropical countries.

A survey of Organic Farming literature by Juroszek et al. (2008b) using the international database CAB-Abstracts showed a lack of research in organic vegetables compared to other organically produced plant products. Only 70 records for the four most important vegetable crops (tomato, lettuce, carrot, cucumber) were returned compared to 208 records for the four most important field crops (wheat, potato, rice, corn), 191 records for the four most important fruit crops (apple, strawberry, grape, banana) and even 142 records for the four most important luxury foodstuffs including coffee, wine, tea, and beer (Juroszek et al., 2008b). Although in several publications and agricultural statistics potato is considered to be a vegetable crop, however, in Juroszek et al. (2008b) and in this paper potato is considered to be a field (or arable) crop.

The limited literature returned showed that tomato is the vegetable most commonly researched in Organic Farming systems, followed by lettuce, carrot and cucumber (Table 1). If potato would have been considered to be a vegetable, tomato would be the second most important vegetable crop under research in organic vegetable production.

A search of CAB-Abstracts (Table 1), shows that organic vegetable research is only a small subset of overall vegetable research – most of which has been done in developed countries under temperate and subtropical climatic conditions. The FAO document "Organic Agriculture and Food Security" also states that there is almost no organic agricultural research taking place in most countries and current allocations in developed countries do not exceed one percent of total agricultural research budgets (Scialabba EL-Hage, 2007). For example, "organic" tomato research was mainly carried out in three countries including the USA (9 records returned), Italy (6 records returned), and Turkey (5 records returned). Only four records were found on "organic tomato" production in tropical countries – three from Brazil and one from India. Results on many other vegetable species such as radish, fennel, celery, chili pepper or basil are not available at all in refereed journals and conference proceedings (Table 1). These observations are in agreement with a recently published review paper by Badgley et al. (2007), who compared yields of organic versus conventionally produced food worldwide.

Table 1: Number of records returned when entering specific keywords related to organic vegetable production

Key words entered (scientific names are available in Annex 1)	No. of records returned CAB-Abstracts + (AGRICOLA, AGRIS) [1)]	No. of records (in front of the letter) related to continents where the organic research activity was conducted	No. of records returned CAB-Abstracts only[2)] (without 'organic' in front, vegetable name only)
Organic tomato	27 + (9) = 36	9AS, 4LA, 9NA, 14E	32,536
Organic lettuce	16 + (3) = 19	3LA, 4NA, 12E	7,255
Organic carrot	9 + (2) = 11	2O, 9E	5,684
Organic cucumber	8 + (1) = 9	3AS, 2LA, 1NA, 3E	11,411
Organic onion	7 + (1) = 8	1LA, 1NA, 5E, 1O	7,349
Organic pepper	5 + (0) = 5	3AS, 2NA	9,026
Organic bean	5 + (0) = 5	1LA, 1NA, 3E	25,192
Organic pea	5 + (0) = 5	4E, 1O	16,740
Organic cabbage	5 + (0) = 5	2LA, 1NA, 2E	9,374
Organic broccoli	4 + (1) = 5	1LA, 4NA	2,847
Organic cauliflower	3 + (0) = 3	3E	4,457
Organic garlic	2 + (1) = 3	1AF, 2LA	5,253
Organic spinach	2 + (1) = 3	1NA, 2E	5,926
Organic asparagus	0 + (2) = 2	2NA	2,828
Organic sweet corn	1 + (1) = 2	2NA	1,043
Organic zucchini	0 + (1) = 1	1AF	1,185
Organic leek	1 + (1) = 2	2E	1,171
Organic eggplant	1 + (0) = 1	1LA	1,925
Organic chicory	0 + (1) = 1	1E	1,735
Organic Brussels sprout	1 + (0) = 1	1E	114
Organic bell pepper	1 + (0) = 1	1NA	730
Organic chilli/chili	0	-	186/86
Organic celery	0	-	1,622
Organic fennel	0	-	1,090
Organic red beet	0	-	438
Organic kohlrabi	0	-	415
Organic radish	0	-	3,884
Organic pumpkin	0	-	1,678
Organic basil	0	-	908

(Databases AGRICOLA 1984 – September 2007, CAB-Abstracts 1989 – September 2007, AGRIS 1975 – June 2007, accessed 8 November 2007)

[1)] The same records already returned in CAB-Abstracts not included to avoid duplicate records

[2)] The database CAB-Abstracts was used only in order to avoid duplicate records

Note: **AF** = Africa (mostly countries of Northern Africa), **AS** = Asia (including Turkey, Israel, and other countries of the Near East), **E** = Europe, **LA** = Latin America (most records refer to Brazil), **NA** = North America (all records refer to USA except one that refers to Canada), **O** = Oceania (records refer to Australia and New Zealand only); note: a few more records may have been returned if synonyms for 'organic' such as 'ecological', 'nature', and 'alternative' were also used.

Research topics under investigation and their relative importance

The total number of research records on organic vegetable production is small and the range of topics covered is large (Table 2). In lettuce, microbial contamination from animal manures was important, as was weed control in organic onion production (Table 2) and yield comparisons between organic and conventionally grown tomatoes (e.g. Denison et al., 2004). The research needs related to crops that are important in (sub) tropical Asia and Africa such as eggplant (brinjal) have not yet been addressed.

Table 2: Number of publications related to a research topic in organic vegetable production

Vegetable species (scientific names are available in Annex 1)	Number of records returned (see Table 1)	Number of publications (in front of the letter) related to a research topic according to the title of the publication (abbreviations used see bottom of Table)
Organic tomato	36	**9C**, 5CM, 2GH, 1M, 7N, 1Q, 7P, 2PH, 2V
Organic lettuce	19	1CM, 2GH, **6H**, 3M, 1Q, 5P, 1PH
Organic carrot	11	1C, 1CM, 1M, **5P**, 3W
Organic cucumber	9	1C, **3GH, 3P**, 1PH, 1V
Organic onion	8	1CM, 1M, 1N, 1V, **4W**
Organic pepper	5	1 CM, 1N, **3P**
Organic bean	5	**2N, 2CM**, 1P
Organic pea	5	2P, **3W**
Organic cabbage	5	1C, **2N, 2P**
Organic broccoli	5	**2M, 2N**, 1P
Organic cauliflower	3	1CM, 1N, 1V
Organic garlic	3	1M, 1N, 1P
Organic spinach	3	2C, 1Q
Organic asparagus	2	2CM
Organic sweet corn	2	1CM, 1W
Organic zucchini	1	1N
Organic leek	2	1P, 1CM
Organic eggplant	1	1N
Organic chicory	1	1CM
Organic Brussels sprout	1	1P
Organic bell pepper	1	1CM

Abbreviations used: **C** = Comparisons of production systems (organic vs. conventional in regard to different aspects, many are related to the LTRAS long-term experiment in California, USA); **CM** = Crop Management (various aspects of crop management, production guides, etc.); **GH** = Green-House production in general (mostly in combination with fertilizer application treatments); **H** = Hygiene (related to contamination of vegetables with harmful microorganisms via e.g. application of animal manures, often compared with conventionally produced lettuce); **M** = Marketing (including other economic aspects); **N** = Nutrition (including soil, green manures, mulching, etc.); **Q** = Quality of products (including food safety except microbial contamination); **P** = Plant protection (diseases, insect pests, and nematodes, beneficial organisms for biological control, etc.); **PH** = Post-Harvest (including shelf life, drying of products, etc.); **V** = Variety evaluation; **W** = Weed management;

Institutionalized research in organic vegetable production in most (sub) tropical countries is either nonexistent or very new. The same is true for most of the International Agriculture Research Centers with the exception of AVRDC-The World Vegetable Center which established an Organic Farming research programme in 2004. The need for more international research was recently recommended by FAO, which called for a "Consultative Group for Organic Agricultural Research entrusted to develop the organic research agenda within the CGIAR system and with national research institutions" (Scialabba EL-Hage, 2007). In addition to the CGIAR system and the United Nations through FAO, other international organizations that may help to coordinate institutionalized research approaches in (sub) tropical countries include the International Society of Organic Agriculture Research (ISOFAR), the Global Horticulture Initiative (GlobalHort), and AVRDC-The World Vegetable Center among others. Various international public and private donor agencies may need to financially support a coordinated comprehensive research approach similar to the CGIAR Challenge Program on High-Value Crops - Fruits and Vegetables, in which organic fruit and vegetable production research may deserve to be included as a sub-project.

3.7 Research needs in organic vegetable production in the (sub) tropics

According to an electronic survey by Stoll (2003) tropical lowland vegetable production is the predominant form of organic vegetable cultivation in the tropics followed by upland vegetable cultivation. The major crop production constraints faced include various pest impacts, poor soil fertility, and plant nutrition. Other important constraints include a lack of superior varieties and high-quality seeds, and poor linkages between vegetable producers and their input suppliers and produce markets.

Relative importance of research needs

Managing pests such as weeds, nematodes, insect pests, mites, and diseases appear to be greater problems for organic vegetable farmers in the (sub) tropics than managing soil fertility. Such problems and related research needs are also influenced by the vegetable species and varieties grown, and by the seasonal and site-specific production conditions.

According to the volume '*Organic Vegetable Production in Malaysia*' (Aini et al., 2005) pest and pathogens are the major impediments to the development of organic vegetable farming in Malaysia, and the cause of some farmers ceasing organic vegetable cultivation after only a few seasons. This agrees with the experiences of experts in IPM Farmer Field Schools, who claimed that developing alternative pest management options is the biggest challenge in organic vegetable production (Stoll, 2003). It appears to be easier for farmers in the tropics to maintain or increase their soil fertility and provide appropriate crop nutrition levels compared to solving plant protection problems. There are many organic options for meeting nutrient demands including green manures, animal manures, extracts of fish residues, home-made plant enzyme extracts, plant residues and composts (Lathiff & Maraikar, 2003; Aini et al., 2005). Research is still needed to ensure the most effective use of limited nutritional resources such as phosphorus or to provide the best management options for stressed soils affected by salinity, acidification, nutrient depletion or low organic matter content (e.g. Grenz & Sauerborn, 2007; Anderson et al., 2007).

Research needs – key insect pests

Only general statements about the most troublesome and economically important insect pests can be made, as few of the respondents to Stoll's survey (2003) specified the pests involved. The pest complex of crucifers was regarded as a major problem; including flea beetle (*Phyllotreta spp.*), cabbage webworm (*Hellula undalis*), and diamondback moth (*Plutella xylostella*). In eggplant, a key insect pest is the eggplant shoot and fruit borer (*Leucinodes orbonalis*). In various other vegetable crops insect pests such as aphids (e.g. *Myzus persicae, Aphis gossypii*), mites (e.g. broad mite *Polyphagotarsonemus latus*), tomato fruitworm (*Helicoverpa armigera*), beet armyworm (*Spodoptera exigua*), and cutworm (*Spodoptera litura*) were mentioned as key insect pests. Soil-borne diseases and insect pests are in general very difficult to manage (SP-IPM, 2004). Our experiences with on-farm and on-station research in organic vegetable production in Taiwan under (sub) tropical conditions suggests that plant virus vectors such as whiteflies (e.g. *Bemisia tabaci*) and melon fruit fly (*Dacus cucurbitae*), which reduce the marketable fruit yield of cucurbits are also key insect pests (Juroszek et al., 2007). In general, we found that it is more difficult to manage small sucking insect pests such as whiteflies in the open field without protective structures such as nets and screen houses to exclude them. In contrast, larvae of several economically important *Lepidoptera such as Spodoptera species* can be controlled relative easily with *Bacillus thuringiensis* (*Bt*) foliar applications as long as larvae are small and remain exposed on leaves and stems. A number of organic plant protection measures for various crops in the (sub) tropics are now well established (e.g. UN, 2003; Aini et al., 2005; OISAT, 2007).

Research needs – key plant diseases

The best means of disease control involves using resistant varieties. Fungal disease control is in general difficult in Organic Farming due to a lack of effective fungicides (Aini et al., 2005). Respondents to Stoll's survey (2003) claimed that in tomato, late blight (*Phytophtora infestans*) is a major concern. In many tropical countries, bacterial wilt (*Ralstonia solanacearum*) and virus diseases (eg. Tomato Yellow Leaf Curl Virus) are also major constraints in crops such as tomato, and only limited control of such viruses is possible by managing their vectors such as whiteflies (AVRDC, 2007a). The most important research need in organic and conventional farming is the development of disease-resistant or tolerant varieties (Aini et al., 2005). Mechanical controls such as grafting also can be considered to manage soil-borne diseases for economically important crops such as cucumber and tomato (George & Eghbal, 2003). Field experiments in tropical Asian countries have shown that grafting tomato onto eggplant rootstocks increased tomato tolerance to flooding and bacterial wilt disease compared to non-grafted tomato plants (Palada & Wu, 2007).

Research needs - botanical pesticide development

In the last decade there has been a great increase in research papers and books published on natural pesticide development and use (e.g. Prakash & Rao, 1997; Renault-Roger et al., 2005). Only a few botanical pesticides such as garlic, hot pepper, pyrethrum, rotenone, and particularly neem products from the tree *Azardirachta indica* have been intensively investigated and commercialized (Isman, 2006). More than 1,000 plants with potential applications in crop protection have yet to be investigated in detail (Prakash & Rao, 1997). Field experiments are needed to confirm the abilities of the most promising candidates observed in laboratory studies (Kaushik, 2004).

Commercial and homemade botanical pesticides may be of most benefit in (sub) tropical developing countries where synthetic pesticides are unavailable or too costly, and where traditions of using plant derivatives for the protection of crops and stored food exists (Isman,

2006). Even where synthetic pesticides are available and affordable to growers, a lack of protective equipment and the negative effect of these substances on the environment can justify the development and use of effective, less harmful homemade botanical pesticides (Schuhbeck & Bokosou, 2001). Many recipes for homemade botanicals can be found in Stoll (2005). However, the efficacy, human and environmental toxicity, and mode of action of most homemade botanical pesticides remains to be systematically studied. Research should consider aspects influencing their efficacy, such as optimized extraction methods, the use of natural additives such as alcohol, storage conditions, target pests in relation to their vegetable host plants, and phytotoxicity.

A combination of preventive and direct plant protection methods is the most successful strategy in organic vegetable production (Varghese, 2000; Oad & Siddiqui, 2004; Treadwell, 2007). In general, preventive methods such as maintaining soil fertility and health are geared towards enhancing the vigor and health of crop plants (UN, 2003).

Research needs – increased crop biodiversity

Most conventional and organic farmers in developed and developing countries grow only a small number of vegetable species including tomato, pepper, cabbage, onion, and cucumber, because there are good marketing opportunities for these crops. There are many more options available. Information on about 350 African indigenous vegetables and approximately 7,000 other useful plants of tropical Africa is summarized in the Plant Resources of Tropical Africa database (PROTA Foundation, 2007). Greater use of often neglected indigenous vegetables can lead to more diversity in diets, less seasonality in vegetable production and increased farmers' income, as well as the preservation of genetic diversity (Engle, 2003). According to Adebooye & Opabode (2004), indigenous species are usually better adapted to local conditions, and resistant to local pests and diseases. Some also grow well under low-fertility situations (Keller, 2004). Indigenous vegetables can become a useful part of local cropping systems if there is a market. For instance, indigenous vegetables such as the African eggplant (*Solanum aethiopicum*, *S. macrocarpon*, *S. anguivi*) are popular across much of Africa (AVRDC, 2004) and in Tanzania (Weinberger & Msuya, 2004) where it has become a cash crop. However, much less research has been conducted on indigenous vegetables compared to the important vegetable crops such as tomato. The European Commission-funded project 'IndigenoVeg' supports networking activities among researchers and aims at identifying the research needs in African indigenous vegetables.

3.8 Vegetable variety screening and breeding for low-input and low soil fertility conditions with a focus on phosphorous (P) efficiency

A critical issue for vegetable production in the (sub) tropics is the issue of soil nutrient mining. Vegetable production requires moderate to high inputs of plant nutrients to achieve economic yields and quality, but most "organic by default" farmers in sub-Saharan Africa are farming depleted soils. In addition, there is an impending global shortage of high-grade sources of phosphate fertilizers. Therefore, the adaptation of agricultural production systems such as cereal and vegetable farming to low phosphate conditions will be increasingly important for both organic and conventional farmers.

Fertilizer use in sub-Saharan Africa is very low, and soil depletion and degradation are global problems. The total average annual mineral fertilizer supply in sub-Saharan Africa is roughly about 12 kg N, 5 kg P, and 2 kg K per ha of cropland, and this is unevenly distributed (Grenz & Sauerborn, 2007). Worldwide, an estimated 2 billion hectares of cropland are degraded from erosion, crusting, sealing, acidification, salinity, pollution, organic matter depletion, loss of soil flora and fauna, and nutrient depletion (SP-IPM, 2004).

Efficient P-uptake and use by plants is already very important in a range of different production systems in areas with naturally low P-fertility, such as in Africa or Australia (Grenz & Sauerborn, 2007). Being able to produce effectively under low fertility systems may become more common (Juroszek et al., 2008b) due to expanding land degradation and soil depletion as well as the expected depletion of high-grade phosphate fertilizer sources this century (Runge-Metzger, 1995). Marked changes in root morphology and architecture under P-deficiency can help plants to explore the soil for P and many plants have developed a symbiotic relationship with mycorrhizal fungi to increase their P-availability (Raghothama & Karthikeyan, 2005). Varietal differences may be considered in breeding programmes.

However, breeding of modern varieties is conducted mostly under high-input situations and breeding has systematically missed out on exploiting genetic differences expressed at low levels of inputs (Ceccarelli, 1996). Compared with cereals, there is less information available in the literature on vegetable variety evaluation under low-input and low soil fertility systems, especially under (sub) tropical conditions. An exception is the study by De Melo (2003) conducted in Central Europe focusing on root traits contributing to a better adaptation of onion and its allied species *Allium fistulosum* to organic systems.

Baresel et al. (2005) pointed out that for very low-yield environments, selection in low-input systems is necessary, and this can be successfully done (El-Bassam, 1998; Lammerts van Bueren, 2002). For instance Moura et al. (2001) using a pot experiment, found a large variation in the phosphorus (P) efficiency of 10 tested sweet pepper lines, suggesting that genetic improvement programmes aimed at increasing these characteristics in sweet pepper could be successful. The sweet pepper lines differed mainly in rooting, P-uptake and P-use efficiency. In contrast, only small differences in the P-translocation efficiency were observed among lines.

To cope with expanding low-fertility situations, other approaches in addition to breeding P-efficient vegetable varieties should be considered. These may include the use of microbes for enhancing P-availability, improved technologies to use low-grade phosphate rocks (Raghothama & Karthikeyan, 2005), and extraction and recycling of 'safe' P from organic resources such as waste water and solid waste (Grenz & Sauerborn, 2007).

3.9 Conclusions

Research on organic vegetable production in the (sub) tropics has not yet been conducted intensively. It is a new area of research both for national and international agricultural research centres. However, improving the performance of vegetable production systems through research and development is important for organic and conventional vegetable farming alike. In sub-Saharan Africa the majority of farmers are "organic by default" and these farmers are also likely to be most at risk from the impact of climate change (CGIAR, 2006). With increasing safety concerns from misuse of pesticides and expected increases in the cost of some key mineral inputs such as phosphate fertilizer, many of the principles of successful Organic Farming that could be clarified through research will also be of importance to conventional farming.

Research needs specifically related to Organic Farming technologies include the identification and development of effective alternative plant protection and plant nutrition approaches. Research important for both organic and conventional vegetable production systems includes the breeding of superior vegetable varieties that are resistant to various pests and pathogens and tolerant to abiotic stressors including heat, salinity, drought, temporary flooding and low soil fertility. Breeding work needs to be done under conditions of low fertility as well as high fertility. Little work has been done to breed vegetable crop varieties with efficient P-uptake and use. Breeding has systematically missed out on exploiting genetic differences expressed at low levels of inputs. Shuttle breeding techniques using contrasting environments and the multilocation international testing approach practiced by Borlaug (2007) in cereals could be useful in vegetables as well. If the contrasting environments include organic as well as high-input and low-input conventional farming systems, it could make the outcome of agricultural research more robust for adaptation to a wider range of cropping systems and environmental conditions.

Acknowledgments
We wish to thank the donors who support AVRDC's Organic Vegetable Program including BMZ/GTZ, Germany; The Organic Center for Education & Promotion, USA; and the Council of Agriculture, Taiwan. We are grateful to Mrs Maureen Mecozzi for editing the manuscript.

References

Abele, S.; Dubois, T.; Twine, E.; Sonder, K. & Coulibaly, O. (2007) Organic agriculture in Africa: A critical review from a multidisciplinary perspective. Hülsebusch, C.; Wichern, F.; Hemann, H. & Wolff, P. (eds.) Organic agriculture in the Tropics and Subtropics – Current status and perspectives. Supplement to the Journal of Agriculture and Rural Development in the Tropics and Subtropics, kassel university press GmbH, 143-166, <http://www.upress.uni-kassel.de/online/frei/978-3-89958-263-volltext.frei.pdf>

Adebooye, O.C. & Opabode, J.T. (2004) Status of conservation of the indigenous leaf vegetables and fruits of Africa. African Journal of Biotechnology **3**, 700-705

Aini, Z.; Sivapragasam, A.; Vimala, P. & Mohamad Roff, M.N. (2005) Organic Vegetable Cultivation in Malaysia. Malaysian Agriculture Research and Development Institute (MARDI). Kuala Lumpur

Anderson, A.; Kelly, J. & McKenzie, D. (eds.) (2007) Healthy Soils for Sustainable Vegetable Farms: Ute Guide. Land & Water Australia and AUSVEG Ltd, Arris Pty Ltd.

AVRDC (2004) AVRDC Report 2003. AVRDC-The World Vegetable Center, Shanhua, Taiwan. <http://www.avrdc.org/publications/annual_report/AR_2003.pdf>

AVRDC The World Vegetable Center (2007a) Extension Materials. <http://www.avrdc.org/LC/tomato/publications.html>

AVRDC The World Vegetable Center (2007b) Extension Materials. http://www.avrdc.org/LC/home.html

Badgley, C.; Moghtader, J.; Quintero, E.; Zakem, E.; Chappell, M.J.; Aviles-Vazquez, K.; Samulon, A. & Perfecto, I. (2007) Organic agriculture and the global food supply. Renewable Agriculture and Food Systems **22**, 86-108

Barker, R. (1979) Adoption and Production Impact of New Rice Technology – The Yield Constraints Problem. IRRI (1979) Farm-level constraints to High Rice Yields in Asia: 1974-77. International Rice Research Institute, Los Banos, Laguna, Philippines, 1-27

Baresel, J.P.; Reents, H.J. & Zimmermann, G. (2005) Field evaluation criteria for nitrogen uptake and nitrogen use efficiency. Proc. COST SUSVAR/ECO-PB Workshop on organic plant breeding strategies and the use of molecular markers. Driebergen, The Netherlands, 17-19 January, 49-54

Bellon, S.; De Abreu, L.S. & Valarini, P.J. (2005) Relationships between social forms of organic horticultural production and indicators of environmental quality: a multidimensional approach in Brazil. Proc. First Scientific Conference of the International Society of Organic Agriculture Research (ISOFAR). Adelaide, Australia, 21-23 September, 430-433

Bett, K.E. & Freyer, B. (2007) Recognizing and realizing the potential of organic agriculture in Kenya. 9. Wissenschaftstagung Ökologischer Landbau <http://orgprints.org/9816/01/9816_Bett_Poster.pdf>

Borght Van der, D.; Litthamalay, S. & Khamphouvong, P. (2004) The Path to pesticides...? A Case Study on trends and tendencies in the Lao PDR. FAO Inter-Country Programme for the IPM in Vegetables in South and Southeast Asia (GCP/RAS/191/AUL)

Borlaug, N.E. (2007) Sixty-two years of fighting hunger: personal recollections. Euphytica **157**, 287-297

Brandt, K. & Kidmose, U. (2002) Nutritional Consequences of using organic agricultural methods in developing countries. Cakmak, I.; Graham, R.D. & Welch, R.M. (eds) Impacts of Agriculture on Human Health and Nutrition. Encyclopedia of Life Support Systems (EOLSS), Eolss Publishers, Oxford

Caldwell, B.; Rosen, E.B.; Sideman, E.; Shelton, A.M. & Smart, C.D. (2005) Resource Guide for Organic Insect and Disease Management. <http://www.nysaes.cornell.edu/pp/resourceguide/>

Ceccarelli, S. (1996) Adaption to low/high input cultivation. Euphytica **92,** 203-214

CGIAR (2006) Intensified Research Effort Yields Climate-Resilient Agriculture to Blunt Global Warming, Prevent Widespread Hunger <http://www.cgiar.org/newsroom/releases/news.asp?idnews=521>

Community IPM (2006) <http://www.communityipm.org/countries.htm>

Coleman, E. (1995) The New Organic Grower: A Master's Manual of Tools and Techniques for the Home and Market Gardener. 2nd Edition, Chelsea Green Publishing Co., White River Junction, Vermont

Danuwat, P.O. (2006) Organic Agriculture in Southeast Asia. <http://info.organic.org.tw/supergood/ezcatfiles/organic/img/img/1085/61701466.pdf>

Davies, G. & Lennartsson, M. (2005) Organic Vegetable Production - A Complete Guide. Marlborough, Crowood Press

De Melo, P.E. (2003) The root systems of onion and *Allium fistulosum* in the context of organic farming: a breeding approach. Ph.D. Thesis, Wageningen University, The Netherlands. <http://library.wur.nl/wda/dissertations/dis3468.pdf>

Denison, R.F.; Bryant, D.C. & Kearney, T.E. (2004) Crop yields over the first nine years of LTRAS, a long-term comparison of field crop systems in a Mediterranean climate. Field Crops Research **86**, 267-277

Dinham, B. (2003) Growing vegetables in developing countries for local urban populations and export markets: Problems confronting small-scale producers. Pest Management Science **59**, 575-582

El-Bassam, N. (1998) A concept of selection for "low input" wheat varieties. Euphytica **100**, 95-100

Engle, L.M. (2003) Collection and conservation of indigenous vegetable germplasm to enhance biodiversity and maintain livelihoods in ASEAN. Perspectives of ASEAN Cooperation in Vegetable Research and Development. Proc. on the ASEAN-AVRDC Regional Network on Vegetable Research and Development (AARNET), Shanhua, Taiwan 24-26 September, p. 65-77. <http://www.avrdc.org/pdf/aarnet_proc/003.pdf>

EU Commission of the European Union (2005) Monitoring of Pesticide residues in Products of Plant Origin in the European Union, Norway, Iceland and Liechtenstein <http://ec.europa.eu/food/fvo/specialreports/pesticide_residues/report_2003_en.pdf>

FAO/WHO (2005) The Codex Alimentarius Commission and the FAO/WHO Food Standards Programme. Guidelines for the Production, Processing, Marketing, and Labelling of Organically Produced Foods <ftp://ftp.fao.org/codex/Publications/Booklets/Organics/organic_2005e.pdf>

FAO (2007a) Statistical Yearbooks 2005-2006 Issue 1. <http://www.fao.org/statistics/yearbook/vol_1_1/pdf/b03.pdf>

FAO (2007b) Publications related to commodities. http://www.fao.org/organicag/simplesearch.do?main_category=601&type=docs&descr=Commodities>

Freidberg, S. (2003) Cleaning up down South: supermarkets, ethical trade and African Horticulture. Social and Cultural Geography **4**, 27-43

Freyer, B. (2007) Traditional and ecological farming systems in (sub) tropical countries – history, interactions and future perspectives. Hülsebusch, C.; Wichern, F.; Hemann, H. & Wolff, P. (eds.) Organic agriculture in the Tropics and Subtropics – Current status and perspectives. Supplement to the Journal of Agriculture and Rural Development in the Tropics and Subtropics, kassel university press GmbH, 1-30 <http://www.upress.uni-kassel.de/online/frei/978-3-89958-263-5.volltext.frei.pdf>

Garrett H.J. & Beck C.M. (1999) Texas Organic Vegetable Gardening. Gulf Publishing Company, Houston, Texas

Geisen, J. (1999) The sustainable vegetable production project in Guimaras – empowering local farmers to increase farm income through environmental-friendly vegetable production. Tropenlandwirt **100**, 99-111

Gennaro L. & Quaglia, G.B. (2003) Food safety and nutritional quality of organic vegetables. Acta Horticulturae **614**, 675-680 <http://www.diis.dk/graphics/Publications/WP2007/WP2007-3%20til%20web.pdf>

George, E. & Eghbal, R. (eds.) (2003) Ökologischer Gemüsebau - Handbuch für Beratung und Praxis. Bioland Verlags GmbH, Mainz. In German

Global Horticulture Assessment (2005) <http://caes.ucdavis.edu/IntProg/GHAPublication2005Web.pdf>

Gonzalez, M.; Miglioranza, K.S.B.; de Moreno, J.E.A & Moreno, V.J. (2005) Evaluation of conventionally and organically produced vegetables for high lipophilic organochlorine pesticide (OCP) residues. Food and Chemical Toxicology **43**, 261-269

Grenz, J.H. & Sauerborn, J. (2007) The potential of organic agriculture to contribute to sustainable crop production and food security in Sub-Saharan Africa. Hülsebusch, C.; Wichern, F.; Hemann, H. & Wolff, P. (eds.) Organic agriculture in the Tropics and Subtropics – Current status and perspectives. Supplement to the Journal of Agriculture and Rural Development in the Tropics and Subtropics, kassel university press GmbH, p. 50-84. <http://www.upress.uni-kassel.de/online/frei/978-3-89958-263-5.volltext.frei.pdf>

Hamm, U. (2005) How to build up international statistics on organic markets. Proc. First Scientific Conference of the International Society of Organic Agriculture Research (ISOFAR). Adelaide, Australia, 21-23 September, 356-359

IFAD International Fund for Agricultural Development (2005) Organic agriculture and poverty reduction in Asia. China and India Focus. Report No. 1664 <http://www.ifad.org/evaluation/public_html/eksyst/doc/thematic/organic/asia.pdf>

IFOAM International Federation of Organic Agriculture Movements (2004) The Gene Revolution: No potential for the poor – yet a threat to the environment <http://www.ifoam.org/press/press/gene_revolution.html>

IPM DANIDA (2007) <http://thailand.ipm-info.org/ipmdanida/ipmdanida.htm>

Isman, M.B. (2006) Botanical insecticides, deterrents, and repellents in modern agriculture and an increasingly regulated world. Annu. Rev. Entomol. **51**, 45-66

Juroszek, P.; Ma, C.H.; Tsai, H.H.; Wu, D.L. & Palada, M.C. (2007) Organic farming research at AVRDC – The World Vegetable Center: developing systems for smallholder farmers in the tropics. Tropentag 2007, Witzenhausen, Germany, 9-11 October, 9 <http://www.tropentag.de/2007/abstracts/full/360.pdf>

Juroszek, P.; Ledesma, D.; Ma, C.H.; Yang, R.Y.; Lumpkin, H.M.; Lin, C.C.; Tsai, H.H.; Wu, D.L.; Hanson, P.M. & Palada, M.C. (2008a) Plant vigor and yields of organically and conventionally grown tomato crops in Taiwan. Acta Horticulturae, in press

Juroszek, P.; Lumpkin, T.A. & Palada, M.C. (2008b) Sustainable vegetable production systems. Acta Horticulturae, in press

Kaffka, S.; Bryant, D. & Denison, F. (2005) Comparison of organic and conventional maize and tomato cropping systems from a long-term experiment in California. Proc. First Scientific Conference of the International Society of Organic Agriculture Research (ISOFAR). Adelaide, Australia, 21-23 September, 218-221

Kaushik, N. (2004) Biopesticides for sustainable agriculture – prospects and constraints. TERI Press, New Delhi

Keller, G.B. (2004) African nightshade, eggplant, spiderflower et al. – production and consumption aspects of traditional vegetables in Tanzania from the farmers point of view. Göttingen, Georg-August Universität, MSc Thesis

Kilcher, L. (2007) How organic agriculture contributes to sustainable development. Hülsebusch, C.; Wichern, F.; Hemann, H. & Wolff, P. (eds.) Organic agriculture in the Tropics and Subtropics – Current status and perspectives. Supplement to the Journal of Agriculture and Rural Development in the Tropics and Subtropics, kassel university press GmbH, 31-49. <http://www.upress.uni-kassel.de/online/frei/978-3-89958-263-5.volltext.frei.pdf>

Koh, D. & Jeyaratnam, J. (1996) Pesticides hazards in developing countries. The Science of the Total Environment **188**, S78-S85

Köpke, U. (2003) Influence of organic and conventional farming systems on nutritional quality of food. Cakmak, I. & Welch, R.M. (eds.), Impacts of Agriculture on Human Health and Nutrition, Encyclopedia of Life Support Systems (EOLSS), Eolss Publishers, Oxford

Kristiansen, P. (2006) Overview of organic agriculture, 1-23. Kristiansen, P.; Taji, A. & Reganold, J. (eds.) Organic Agriculture - A Global Perspective. CSIRO Publishing, Collingwood Victoria

Lammerts van Bueren, E.T. (2002) Organic plant breeding and propagation: Concepts and strategies. PhD Thesis Wageningen University, The Netherlands. <http://www.louisbolk.org/downloads/1310.pdf>

Lathiff, M.A. & Maraikar, S. (2003) Studies on the performance of some vegetable crops in organic farming system. Annals of the Sri Lanka Department of Agriculture **5**, 141-148

Luedeling, E. & Wichern, F. (2007) The need for site-specific adaptations of organic standards: The example of dryland salinity in Australia. Hülsebusch, C.; Wichern, F.; Hemann, H. & Wolff, P. (eds.) Organic agriculture in the Tropics and Subtropics – Current status and perspectives. Supplement to the Journal of Agriculture and Rural Development in the Tropics and Subtropics, kassel university press GmbH, 121-142 <http://www.upress.uni-kassel.de/online/frei/978-3-89958-263-5.volltext.frei.pdf>

Mäder, P.; Fliessbach, A.; Dubois, D.; Gunst, L.; Fried, P. & Niggli, U. (2002) Soil fertility and biodiversity in organic farming. Science **296**, 1694-1697

Mausch, K.; Mithoefer, D.; Asfaw, S. & Waibel, H. (2006) Impact of EurepGAP Standard in Kenya: Comparing smallholders to large-scale vegetable producers. <http://www.tropentag.de/2006/abstracts/full/253.pdf>

Moura, W.M.; Lima, P.C.; Casali, V.W.D.; Pereira, P.R.G. & Cruz, C.C. (2001) Phosphorus efficiency of sweet pepper lines. Horticultura-Brasileira **19**, 306-312. In Portuguese with English abstract

Nath, P. (2007) Cucurbits – Everyone's crop. Acta Horticulturae **731**, 485-491

Oad, F.C. & Siddiqui, M.H. (2004) Organic vegetable cultivation. Indus Journal of Biological Sciences **1**, 124-129

OISAT (2007) On-line Information Service for Non-Chemical Pest Management in the Tropics. <http://oisat.org/controlmap.htm>

Organic Trade Association (2006) U.S. Organic industry overview. OTA's 2006 manufacturer survey. Greenfield, Massachusetts, Organic Trade Association

Palada, M.C.; Kalb, T.J. & Lumpkin, T.A. (2006) The role of AVRDC-The World Vegetable Center in enhancing and promoting vegetable production in the tropics. Hort Science **41**, 556-560

Palada, M.C. & Wu, D.L. (2007) Increasing off-season tomato production using grafting technology for peri-urban agriculture in Southeast Asia. Acta Horticulturae **742**, 125-131

Pai, C.C. (2006) Organic farming in Taiwan <http://info.organic.org.tw/supergood/ezcatfiles/organic/img/img/1085/246923954.pdf>

Parrott, N.; Ssekyewa, C.; Makunike, C. & Ntambi, S. (2006) Organic Farming in Africa. Willer, H. & Yussefi, M. (eds.) (2006) The World of Organic Agriculture. Statistics and Emerging Trends 2006. IFOAM (Bonn, Germany) and FiBL (Frick, Switzerland)

Peet, M. (1996) Sustainable Practices for Vegetable Production in the South. <http://www.cals.ncsu.edu/sustainable/peet/>

Peng, S.; Cassman, K.G.; Virmani, S.S.; Sheehy, J. & Khush, G.S. (1999) Yield potential trends of tropical rice since release of IR8 and the challenge of increasing rice yield potential. Crop Science **39**, 1552-1559

Pimentel, D.; Hepperly, P.; Hanson, J.; Douds, D. & Seidel, R. (2005) Environmental, energetic, and economic comparisons of organic and conventional farming systems. BioScience **55,** 573-582

Prakash, A. & Rao, J. (1997) Botanical Pesticides in Agriculture. CRC Press, Lewis Publishers, New York

PROTA Foundation (2007) Plant Resources of Tropical Africa. <http://www.prota.org/uk/About+PROTA/>

Raghothama, K.G. & Karthikeyan, A.S. (2005) Phosphate acquisition. Plant and Soil **274**, 37-49

Rajotte, E.G.; Norton, G.W.; Luther, G.C.; Barrera, V. & Heong, K.L. (2005) IPM Transfer and Adoption, 143-157. Norton, G.W.; Heinrichs, E.A; Luther, C.G. & Irwin, M.E. (eds.) Globalizing Integrated Pest Management – A Participatory Research Process, Blackwell Publishing, Ames, Iowa

Ramesh, P.; Singh, M. & Rao, A.S. (2005) Organic farming: its relevance to the Indian context. Current Science **88,** 561-568

Raupp, J.; Pekrun, C.; Oltmanns, M. & Köpke, U. (eds.) (2006) Long Term Field Experiments in Organic Farming. ISOFAR Scientific Series No 1, Verlag Dr. Köster, Berlin

Regnault-Roger, C.; Philogene, B.J.R. & Vincent, C. (eds.) (2005) Biopesticides of Plant Origin. Lavoisier Publishing, Paris

Roitner-Schobesberger, B.; Darnhofer, I.; Somsook, S. & Vogl, C.R. (2007) Consumer perceptions of organic foods in Bangkok, Thailand. Food Policy, in press

Runge-Metzger, A. (1995) Closing the cycle: Obstacles to efficient P management for improved global food security, 27-42. Thiessen, H. (ed.) Phosphorus in the global environment: Transfers, cycles and management, John Wiley & Sons, New York

Saxena, S. (2007) Organic vegetables: domestic and regional marketing constraints and opportunities for small-scale farmers in East Africa. Tropentag 2007, Witzenhausen, Germany, 9-11 October, 7, <http://www.tropentag.de/2007/abstracts/full/194.pdf>.

Scialabba EL-Hage, N. (2007) Organic Agriculture and Food Security, 3-5 May, FAO, Italy. <http://ftp.fao.org/paia/organicag/ofs/OFS-2007-5.pdf>

Schuhbeck, A. & Bokosou, J. (2001) The potential of using homemade plant-derived pesticides to increase food crop production and local food security. In: Food security for Papua New Guinea, ACIAR Proceedings No. 99, 577-581

Sieling, K.; Stahl, C.; Winkelmann, C. & Christen, O. (2005) Growth and yield of winter wheat in the first 3 years of a monoculture under varying N fertilization in NW Germany. European Journal of Agronomy **22,** 71-84

SP-IPM (2004) Soil biota and sustainable agriculture: challenges and opportunities. IPM Research Brief No. 2. SP-IPM Secretariat. Cotonou, Benin, International Institute of Tropical Agriculture

Stoll, G. (2003) Background study on the current state of organic agriculture research in horticulture and its perspectives in Asia. Report for GTZ/BMZ, AVRDC, and IFOAM

Stoll, G. (2005) Natural Crop Protection in the Tropics, 2nd edition, Margraf Verlag, Weikersheim

Tamm, L.; Köpke, U.; Cohen, Y. & Leifert, C. (2007) Development of strategies to improve quality and safety and reduce cost of production in organic and low input crop production systems. 3rd QLIF Congress, Hohenheim, Germany, 20-23 March 2007, 8. <http://orgprints.org/10626/01/Tamm-etal-2007-crop_production_system.pdf>

Treadwell, D.D. (2007) Introduction to Organic Crop Production. University of Florida, Horticultural Sciences Department, IFAS Extension, Document HS720. <http://edis.ifas.ufl.edu/pdffiles/CV/CV11800.pdf>

UN United Nations (2003) Organic Fruit and Vegetables from the Tropics. United Nations, New York and Geneva. <http://www.unctad.org/en/docs/ditccom20032_en.pdf>

USDA-ARS United States Department of Agriculture-Agriculture Research Service and NAL National Agriculture Library (2003) Organic Gardening: A Guide to Resources. <http://www.nal.usda.gov/afsic/AFSIC_pubs/org_gar.htm>

Varghese, L. (2000) Organic vegetables issues and strategies – a review. Agriculture Review **21**, 211-222

WHO and UNEP (1990) Public health impact of pesticides use in agriculture. WHO Geneva

Weinberger, K. & Msuya, J. (2004) Indigenous Vegetables in Tanzania – Significance and Prospects. Technical Bulletin No. 31, AVRDC Publication No. 04-600. Shanhua, Taiwan, AVRDC-The World Vegetable Center. <http://www.avrdc.org/pdf/TB31.pdf>

Weinberger, K. & Lumpkin, T.A. (2005) Horticulture for poverty alleviation - the unfunded revolution. AVRDC Publication No. 05-613. Shanhua, Taiwan, AVRDC-The World Vegetable Center. <http://www.avrdc.org/pdf/WP15.pdf>

Willer, H. & Yussefi, M. (2007) The World of Organic Agriculture. Statistics and Emerging Trends 2007. IFOAM (Bonn, Germany) and FiBL (Frick, Switzerland)

Winter, C.K. & Davis, S.F. (2006) Organic foods. Journal of Food Science **71**, R117-R124

Woese, K.; Lange, D.; Boess, C. & Bögel, K.W. (1997) A comparison of organically and conventionally grown foods – results of a review of the relevant literature. Journal Sci. Food Agri. **74**, 281-293

Zhang, F.; Qiao, Y.; Wang, F. & Zhang, W. (2007) A perspective on organic agriculture in China – opportunities and challenges. 9. Wissenschaftstagung Ökologischer Landbau. <http://orgprints.org/10388/01/10388_Zhang_Vortrag.pdf>

Zhou, Z. (2003) China's market for organic fruits and vegetables: Current situation and prospects. Proceedings of the Seminar on the Production and Exports of Organic Fruit and Vegetables in Asia. Bangkok, Thailand, 3-5 November 2003. Organized by FAO, IFOAM, and EarthNet Foundation <http://www.fao.org/docrep/006/AD429E/ad429e17.htm#P618_66083>

Zundel, C.; Kilcher, L. & Mäder, P. (2007) What can organic agriculture contribute to sustainable development? – Long-term farming systems comparisons in the tropics. Tropentag 2007, Witzenhausen, Germany, 9-11 October, 4. <http://www.tropentag.de/2007/abstracts/full/491.pdf>

Annex 1. Scientific names of vegetables according to USDA Plants Database

English name	Scientific name (s)
Asparagus	*Asparagus officinalis* L.
Basil	*Ocimum basilicum* L.
Bean	*Phaseolus vulgaris* L. ssp.
Broccoli	*Brassica oleracea* L. var. *italica* Plenck
Brussels sprout	*Brassica oleracea* L. *var. gemmifera* DC
Cabbage	*Brassica oleracea* L. var. *capitata* (may be others as well)
Carrot	*Daucus carota* L. ssp. *sativus* (Hoffmann) Arcang
Cauliflower	*Brassica oleracea* L. var. *botrytis*
Celery	*Apium graveolens* L.
Chicory	*Cichorium intybus* L.
Cucumber	*Cucumis sativus* L.
Eggplant (brinjal)	*Solanum melongena* L.
Fennel	*Foeniculum vulgare* P. Mill.
Garlic	*Allium sativum* L.
Kohlrabi	*Brassica oleracea* L. *var. gongylodes*
Leek	*Allium porrum* L.
Lettuce	*Lactuca sativa* L.
Onion	*Allium cepa* L.
Pepper (bell/sweet)	*Capsicum annuum* L.
Pepper (chilli/chili/hot)	*Capsicum annuum* L.
Pea	*Pisum sativum* L.
Pumpkin	*Cucurbita pepo* L. (may be others as well)
Radish	*Raphanus sativus* L.
Red beet	*Beta vulgaris* L. ssp. *vulgaris*
Spinach	*Spinacia oleracea* L.
Sweet corn	*Zea mays* L.
Tomato	Mainly *Solanum lycopersicum* L. (formerly *Lycopersicon esculentum* Mill.)
Zucchini	*Cucurbita pepo* L.

USDA Plants Database <http://plants.usda.gov/java/nameSearch>

Use of farmyard manure in improving soil physicochemical properties and crop yield in smallholding agriculture in north eastern Ethiopia

Wondimu Bayu

Amhara Region Agricultural Research Institute, P. O. Box 527, Bahir Dar, Ethiopia.
E-Mail: wondimubayu@yahoo.com

4.1 Introduction

In smallholder farming in north eastern Ethiopia several important organic soil amendment practices such as fallowing, crop rotation, recycling of dung and crop residues to the soil, and application of external sources of plant nutrients have been abandoned. The many complicated factors underlying this situation include high population pressure and the resulting small land holding, poverty, high costs of and limited access to agricultural inputs and credit, low profitability of agricultural production, high risks facing farmers, fragmented land holdings and insecure land tenure, and short time horizons of farmers land management. Crop production in north eastern Ethiopia, thus, is characterized by soil organic matter loss and mining of soil fertility due to continuous cereal based cultivation with limited or no application of organic or inorganic fertilizers.

Intensive cropping systems based on cereal growing, intensive soil tillage, and use of mineral fertilizers and pesticides has, of course, substantially increased productivity in the last several years. However, this remarkable productivity increase has not been achieved without cost. The soil resource has gradually lost organic matter contents which have been accompanied by deterioration in soil structure leaving the soil more prone to compaction, erosion, and overall deterioration in physicochemical properties. The continuous removal of biomass from crop land coupled with no input to replenish the soil nutrient and organic matter content have rapidly exhausted the soil nutrient reserves and jeopardized agricultural sustainability. Some soils are now suffering from dangerously low organic matter levels and could not be expected to sustain the farming systems which have been imposed on them. Despite the fact that soil organic matter is a major source of plant nutrients and improves physical properties of the soil, such as soil porosity, structure and water-holding capacity farmers in north eastern Ethiopia pay little attention to maintaining soil organic matter content of their farmland. Furthermore, concerns about soil exhaustion and nutritional imbalances, arising from increased and indiscriminate use of inorganic fertilizers have been raised. Thus, improving soil resources and, thus, food production in the smallholder farm sector is an enormous challenge. Therefore, to increase food production on the existing degraded soils, improving the organic matter content and, thus, the fertility of the soil is critically essential. Several experiments were conducted at different locations in north eastern Ethiopia to study the effects of the integrated use of farmyard manure (FYM) and mineral nitrogen and phosphorus fertilizers on soil properties, crop yield and a parasitic weed (*Striga hermonthica*) infestation. Therefore, this review paper provides evidence on the role of FYM in restoring and maintaining the physicochemical properties and productivity of tropical soils in north eastern Ethiopia.

4.2 Experimental design and site conditions

Two separate experiments, one at Kobo and Chefa and the other at Sirinka were conducted. The experiments at Kobo and Chefa were conducted for six (1997–2002) and four years (1997-2000), respectively on fixed plots, while at Sirinka, it was conducted for three years (1996-1998) on fixed plots. Kobo is situated at 12° 9' N and 39° 38' E at an altitude of 1470 m above sea level with mean annual maximum and minimum air temperatures of 29 °C and 15 °C, respectively and mean annual rainfall of 649 mm (1973–2002), while Chefa is located at 10° 57' N and 39° 47' E and at an elevation of 1450 m a.s.l. with mean annual rainfall of 989.8 mm (35 years) and mean maximum and minimum temperatures of 26 °C and 11.8 °C, respectively. Sirinka research site is situated at 11° 45' N and 39° 36' E at an altitude of 1890 m above sea level with average long term (1980-2002) rainfall of 633 mm and maximum and minimum temperatures of 27.4 °C and 14.3 °C, respectively during the cropping season (July-November). The soil at Kobo is Eutric fluvisol, while at Chefa and Sirinka it is Eutric Vertisol, with the properties summarized in Table 1. At all locations the rainfall is characterized by unpredictability and uneven distribution.

Table 1: Some physicochemical properties of the soils

Soil properties	*Kobo (0-23 cm depth)*	*Chefa (0-20 cm depth)*	*Sirinka (0-30 cm depth)*
pH (1:2.5 in H_2O)	7.5	6.5	6.6
Organic carbon (%)	1.2	2.3	1.5
Total N (%)	0.08	0.18	0.15
available P (Olsen, mg kg^{-1})	12.5	29.4	9.5
K (cmol kg^{-1})	2.3	0.9	ND
Ca^{2+} (cmol kg^{-1})	34.2	27.9	ND
Mg^{2+} (cmol kg^{-1})	11.4	11.2	ND
Na^{+} (cmol kg^{-1})	1.58	0.25	ND
CEC (cmol kg^{-1})	50.6	46.9	ND
Base saturation (%)	ND	87.4	ND
Clay (%)	34	58	59
Silt (%)	52	22	34
Sand (%)	14	20	8

ND = Not Determined

Treatments of the experiments conducted at Kobo and Chefa were comprised of factorial combinations of four rates of FYM (0, 5, 10, and 15 t ha^{-1} on dry-weight basis) and three rates of inorganic fertilizer (0/0, 21/10, and 41/20 kg ha^{-1} N/P) in a randomized complete block design with three replications. Treatments of the experiment conducted at Sirinka were factorial combination of four levels of inorganic nitrogen fertilizer (0, 40, 80, 120 kg ha^{-1} N) and four rates of FYM (0, 10, 20, and 30 t ha^{-1}) arranged in a split-plot design with three replications, where main plot treatments were two crop species, maize and sorghum, and the sub-plot treatments were the N treatments by FYM combinations.

In all cases, FYM was collected from neighbouring farmers' pens, mixed thoroughly, weighed for each plot, spread evenly, and immediately incorporated into the soil a month before planting. The FYM used at Kobo in the six years period contained 0.19-1.38% N,

0.14-0.78% available P, 2.87% K, while the FYM applied at Chefa contained an average of 0.42% N, 0.35% available P, and 0.51% K on dry matter basis. Urea and diammonium phosphate (DAP) at Kobo and Chefa and urea and Triple Super Phosphate (TSP) at Sirinka were used as inorganic fertilizer sources. Half of the urea and all the DAP and TSP were applied in the row at planting and incorporated into the soil. The remainder of the urea was side dressed at the six-to-eight leaf stage. Sorghum cvs "76 T1 #23" at Kobo and Sirinka and "Gambella 1107" at Chefa and maize variety A-511 at Sirinka were hand-drilled in 75 cm rows, later thinned to 15 cm spacing between plants. Gross plot size was 5.25 m wide by 6 m long at Kobo and Chefa and 3.75 m wide by 5 m long at Sirinka. Experimental plots were managed to be free from weeds by hand weeding and insect pest attack through insecticide spraying. At maturity, the central two rows at Kobo and three rows at Chefa and Sirinka were harvested for grain and stover yield determination. Grain yields were adjusted to 12.5% moisture content. At Kobo, plant samples from the sixth year trial were collected at harvest to determine N uptake. The N content of plant samples was determined by the micro-Kjeldahl method (Page et al., 1982). The N uptake in the grain and stover was estimated by multiplying the concentrations with grain and stover yields. Nitrogen-use efficiency was calculated as yield in control plots subtracted from yield in fertilized plots divided by applied inorganic fertilizer N kg ha^{-1} (Yamoah et al., 2002). Data from the 1997 season at Kobo and 1996 at Sirinka were not included, as severe droughts resulted in crop failure. Analyses of variance for stover and grain yields were performed after pooling data over years and testing the homogeneity of error variances using Bartlett's test (Gomez & Gomez, 1984). Soil samples from nine random positions in each plot were taken from the 0-15 cm depth at Kobo and 0-20 cm depth at Chefa and Sirinka prior to planting and FYM application and again after harvest. Soil samples were analyzed for total N, available P, exchangeable cations (Na^+, K^+, Ca^{2+}, Mg^{2+}), organic C, cation exchange capacity, base saturation and pH. Organic carbon was determined by the rapid titration method (Walkley & Black, 1934). Total N and Olsen P were determined according to the procedures outlined by Page et al. (1982). Soil pH was measured in 1:2.5 soil:water suspension. Exchangeable cations (Na^+, K^+, Ca^{2+}, Mg^{2+}) were determined in an ammonium acetate extract at pH 7 using an atomic absorption spectrophotometer. Soil N balance was computed, according to Hegde (1996), as the difference between N added (FYM and fertilizer) and N removed by plants (uptake). Soil water content was determined by the gravimetric method at crop harvest. At Sirinka Striga infestation was counted at weekly intervals from the time of emergence.

Analyses of variance were performed using the SAS statistical program (SAS V8.2, SAS Institute, Inc., Cary, NC) for the Kobo data and MSTATC statistical program (MSTATC, 1989) for the Chefa and Sirinka data. Whenever treatment differences were found significant critical differences were calculated at the 5% level of probability using Duncan's multiple range test.

4.3 Key Results

4.3.1 Soil physicochemical properties

4.3.1.1 Changes in soil chemical properties

Farmyard manure can influence nutrient availability by the total nutrients added, by controlling net mineralization-immobilization process, by serving as source of C and energy to soil microbes, and through interactions with the mineral soil in complexing toxic cations and reducing the P sorption capacity of the soil. In this regard, significant improvements in the soil chemical properties were observed with FYM application at Kobo and Chefa, north eastern Ethiopia. According to Bayu et al. (2006a) and Alemu & Bayu (2005), application of FYM had significantly ($p \leq 0.01$) increased soil total N, available P, and exchangeable K contents by up to 59%, 158%, and 70%, respectively over the control at Kobo (Table 2) and by up to 17 to 24%, 67 to 85%, and 42 to 79%, respectively over the control at Chefa (Table 3). It is also reported that application of FYM also increased Mg^{2+} content by 6 to 30% over the control at Chefa. However, exchangeable cations (Na^{+}, Ca^{2+} and Mg^{2+}), cation exchange capacity, base saturation, and soil pH were not affected by FYM (Tables 2 and 3) both at Kobo and Chefa, except Mg^{2+} at Chefa. Similarly, Powell (1986) reported a three fold increase in soil N and Kaihura et al. (1999) reported a six-fold increase in soil P with FYM applications. Zia et al. (1992) also reported a 12.7% increase in soil total N content with the application of 5 t FYM ha^{-1}.

Table 2: Effect of FYM on soil chemical properties after 6 years of application at Kobo, averaged across inorganic fertilizer treatments

FYM (t ha^{-1})	Total N (%)	P (mg kg^{-1})	K (cmol kg^{-1})	Exchangeable cations (cmol kg^{-1})			Base Saturation (%)	CEC (cmol kg^{-1})	pH
				Na^{+}	Ca^{2+}	Mg^{2+}			
0	0.074	12.9	1.2	0.40	19.02	16.92	76.66	49.11	7.7
5	0.082	19.8	1.7	0.37	16.53	16.44	70.75	49.46	7.6
10	0.093	26.8	1.7	0.39	16.92	15.81	73.97	47.36	7.7
15	0.105	33.4	2.0	0.40	20.45	16.34	79.99	48.86	7.7
s.e.d	0.005	1.55	0.14	0.03	1.99	0.73	4.61	1.64	0.05

Source: Bayu et al. (2006a)

Table 3: Effect of FYM applications on soil chemical properties (0-20 cm depth) at Chefa, averaged across inorganic fertilizer treatments.

FYM (t ha^{-1})	Total N (%)	P (mg kg^{-1})	K (cmol kg^{-1})	Exchangeable cations (cmol kg^{-1})			BS (%)	CEC (cmol kg^{-1})	pH
				Na^{+}	Ca^{2+}	Mg^{2+}			
0	0.17c	26.13b	0.84d	0.25	33.15	10.07b	93.79	49.03	6.6
5	0.18bc	30.72b	1.01c	0.26	33.81	10.70b	94.05	49.58	6.7
10	0.21a	43.75a	1.19b	0.24	33.48	9.99b	96.46	48.67	6.7
15	0.20ab	48.36a	1.50a	0.25	34.02	13.12a	98.10	50.15	6.6
CV (%)	8.9	18.7	7.4	12.9	8.3	14.3	6.6	3.7	2.4

Means within columns followed by the same letters are not significantly different at $p \leq 0.05$.
Source: Alemu & Bayu (2005)

The available evidence indicates that FYM amendments can increase the pH of acid soils and decrease that of calcareous soils (Lungu et al., 1993; Wong et al., 1998; Whalen et al., 2002). Esilaba et al. (2000) in a study conducted at Sirinka also observed a change in soil pH between the samples analyzed before (1996) and after (1997) treatment application (Table 4).

The increase in total soil N with FYM application at Kobo and Chefa is a result of N applied with FYM. While, the increase in available P with FYM addition could be attributed to many factors, such as the addition of P through FYM, the improved availability of soil P as a result of the action of organic acids produced during FYM decomposition, and retardation of soil P fixation by organic anions formed during FYM decomposition (Gill & Meelu, 1982). Increase in soil K may be attributed to the additional K applied through FYM and the solubilizing action of certain organic acids formed during FYM decomposition (Gill & Meelu, 1982).

Table 4: Effect of integrated application of FYM and inorganic N fertilizer on some soil chemical properties in the topsoil (0-20 cm depth) at Sirinka

Inorganic N level (kg ha^{-1})	*FYM level (t ha^{-1})*	*Soil pH*		*Total N (%)*		*Organic carbon (%)*	
		1996	*1997*	*1996*	*1997*	*1996*	*1997*
0	0	6.60	6.75	0.11	0.14	1.60	1.65
0	10	6.83	6.81	0.13	0.14	1.64	1.91
0	20	6.79	6.86	0.12	0.15	1.60	1.95
0	30	6.88	6.94	0.12	0.14	1.64	1.95
40	0	6.84	6.86	0.12	0.12	1.68	1.77
40	10	6.84	6.82	0.12	0.12	1.64	1.83
40	20	6.83	7.02	0.13	0.13	1.75	2.09
40	30	6.88	6.98	0.11	0.13	1.60	2.14
80	0	6.86	6.79	0.13	0.11	1.75	1.81
80	10	6.76	6.89	0.13	0.11	1.75	2.07
80	20	6.88	6.92	0.13	0.13	1.71	1.99
80	30	6.80	6.97	0.18	0.16	1.64	2.07
120	0	6.82	6.73	0.13	0.15	1.79	1.97
120	10	6.81	6.75	0.13	0.13	1.79	1.89
120	20	6.73	6.93	0.13	0.14	1.71	1.99
120	30	6.83	6.86	0.11	0.14	1.64	2.09
Mean		6.81	6.87	0.13	0.13	1.68	1.95
LSD(p<0.05)-Year		0.14		NS		0.22	
-Treatment		NS		NS		NS	

Source: Esilaba et al. (2000)

4.3.1.2 Changes in soil organic carbon

Regular addition of FYM to arable soil either increases soil organic matter content or reduces its rate of loss (Hoffmann et al., 2001). Several researchers (Mokwunye, 1991; Agbenin & Goladi, 1997; Kaihura et al., 1999) have reported an increase in soil organic matter content in the tropics following FYM application. Similarly, in north eastern Ethiopia FYM application has resulted in a highly significant ($p \leq 0.01$) improvement in soil organic carbon content at Kobo and Chefa, but not at Sirinka. Lack of impact on soil organic matter content with FYM application at Sirinka could presumably be due to the short life of the study (Table 4). Increase in soil organic carbon content as high as 15 to 29% at Kobo and 13 to 26% at Chefa were reported with applying 10 and 15 t FYM ha^{-1} (Table 5) (Esilaba et al., 2000; Alemu & Bayu, 2005; Bayu et al., 2006a). The increase in soil organic carbon content with FYM application can be attributed to its stable nature and slow degradability (Goyal et al., 1992). The observed increase in soil organic carbon content in these studies is in agreement with the results of Shirani et al. (2002) who reported a three-fold increase in soil organic matter content with the application of FYM at the rate of 3 and 6 t ha^{-1}. Mokwunye (1991) also reported a two-fold increase with the application of FYM at 5 and 20 t ha^{-1}.

The maintenance of soil organic matter, by adding FYM, has important benefits such as retention and storage of nutrients, increased buffering capacity in low activity clay soils, increased aggregate stability, improved soil macro-structure, improved infiltration, improved water holding capacity, erosion resistance and prevention of soil hardening (Schjonning et al., 1994; Hoffmann et al., 2001).

Table 5: Effect of FYM application on soil organic carbon content at Kobo and Chefa and soil N balance and soil water content at Kobo, averaged across inorganic fertilizer treatments

FYM (t ha^{-1})	Organic carbon (%) Kobo	Organic carbon (%) Chefa	N balance (kg ha^{-1})	Soil water content (%)
0	0.86	3.17c*	-80.7	45.8
5	0.99	3.53bc	-15.7	45.4
10	1.01	3.59b	65.5	46.4
15	1.13	3.98a	141.4	47.4
s.e.d	0.001		5.025	0.534

*Means within columns followed by the same letters are not significantly different at $p \leq 0.05$.
Source: Alemu & Bayu (2005) and Bayu et al. (2006a)

4.3.1.3 Soil N balance

One of the positive contributions of FYM application to farmland is the addition of major plant nutrients. Significant increases in total and NH_4^+- and NO_3^--N contents of the soil with addition of FYM have been reported by several researchers in Africa (Kaihura et al., 1999; Warman & Cooper, 2000; Hoffmann et al., 2001). Similarly, the results of the study conducted in north eastern Ethiopia showed a significant change in soil N balance with FYM applications whereas in plots receiving no FYM and inorganic fertilizer application negative N balance amounting to –99 kg ha^{-1} were recorded (Bayu et al. 2006a). According to the same authors, FYM application at rates of 10 and 15 t ha^{-1} resulted in positive soil N balances with values of 66 kg ha^{-1} and 141 kg ha^{-1}, respectively, while plots with no FYM addition had negative N balance (–81 kg N ha^{-1}) (Table 5). Application of FYM at the lower rate (5 t ha^{-1}) led to a negative N balance (-16 kg ha^{-1}), although it still has improvement of 81% over the plot that received no FYM. In another study in Ethiopia the application of FYM to maize in a hedgerow intercropping system also reversed nutrient balances from net negative balances of –7 to –19 kg N ha^{-1}, – 4 to –12 kg P ha^{-1} and –10 to –26 kg K ha^{-1}, where no nutrients were applied, to net positive balances of 59 to 62 kg N ha^{-1}, 9 to 35 kg P ha^{-1} and 74 to 80 kg K ha^{-1}, where 3 t FYM ha^{-1} were applied (Lupwayi & Haque, 1999).

4.3.1.4 Soil water content

FYM application increases soil water content by increasing the organic matter content of the soil which helps to hold water in the soil. In a study conducted at Kobo Bayu et al. (2006a) reported that FYM application had significantly ($p \leq 0.01$) improved water content of the soil, where applying 10 and 15 t FYM ha^{-1} had increased soil water content by 1.3 and 3.5% over no FYM application (Table 5). The authors attributed the increase in the water content of the soil in manured plots to increased soil organic matter content. Similarly, Powell (1986) reported a 2% greater water holding capacity in manured plots over the control. Carter et al. (1992) also reported 4, 19, and 27% increase in water holding capacity in manured plots over the control.

4.4 Crop yields

4.4.1 Grain yield

By improving the physicochemical properties of the soil, application of FYM often results in improved growth, development and yield of crops. Several studies have reported substantial yield increases from FYM applications (Lungu et al., 1993; Kaihura et al., 1999; Satyanarayana et al., 2002). Likewise, the report by Alemu & Bayu (2005) and Bayu et al. (2006b) showed a significant increase in grain yield at Chefa and Kobo due to the combined application of FYM and inorganic fertilizers. According to these authors, application of 5, 10, and 15 t FYM ha^{-1} along with 21/10 and 41/20 kg ha^{-1} N/P fertilizer rates increased grain yield significantly over the control at Kobo (Table 6), while at Chefa, the highest significant grain yields were obtained with the application of 15 t FYM ha^{-1} with 21/10 and 41/20 kg ha^{-1} N/P fertilizer rate, followed by the application of 10 t ha^{-1} FYM alone (Figure 1). Yield advantages from these applications relative to the control ranged between 14 and 36% at Kobo and 47 and 51% at Chefa.

These results entail that application of 5 t FYM ha^{-1} in combination with 21/10 kg ha^{-1} N/P fertilizer rate resulted in a grain yield comparable to that obtained from applying 41/20 kg ha^{-1} N/P fertilizer rate. Further increase in the FYM rate to 10 and 15 t ha^{-1} and its combination with 21/10 kg ha^{-1} N/P fertilizer rate also resulted in statistically higher yields than applying 41/20 kg ha^{-1} N/P fertilizer alone (100% of the recommended inorganic fertilizer rate). This result suggests that, by applying FYM plus 21/10 kg ha^{-1} N/P fertilizer rate, a cost saving on inorganic fertilizers (worth the price of 20 kg N and 10 kg P ha^{-1}) is possible, with the additional benefit of improving the physicochemical conditions of the soil. Thus, application of FYM has the potential not only to improve crop yield and soil fertility, but also to reduce dependence on inorganic fertilizers. Application of 10 t ha^{-1} FYM alone resulted in N and P fertilizer equivalents higher than 21 kg N ha^{-1} and 10 kg P ha^{-1}, respectively.

According to Bayu et al. (2004), the application of FYM in combination with inorganic N fertilizer had significantly increased maize yield at Sirinka, where application of 80 and 120 kg N ha^{-1} along with 20 and 30 t FYM ha^{-1} gave significantly higher yields, whereas in sorghum the highest grain yields were obtained with the application of 20 and 30 t FYM ha^{-1} (Table 7).

Increases in grain yield with FYM application is ascribed to the addition of soil nutrients and its favourable effect on the general physicochemical and microbial properties of the soil (Lungu et al., 1993; Prasad et al., 2002). Satyanarayana et al. (2002) indicated that FYM plays an important role in improving soil permeability to air and water and water-stable aggregates, thus improving soil properties and nutrient uptake, which in turn results in higher growth and yield. Similar long-term studies on different crops have shown increased yields due to the application of FYM, and these effects were attributed largely to improved soil organic matter, and soil physicochemical and microbial properties (Belay et al., 2001; Satyanarayana et al., 2002). According to Flaig (1974 as cited by Sharma & Mittra, 1988), application of FYM also influences plant growth physiologically by providing growth-regulating substances. Moreover, the combined use of FYM and inorganic fertilizers also checks N losses and may conserve soil N by forming organic-mineral complexes, thus ensuring continuous N availability to the crop (Satyanarayana et al., 2002). Similarly, increase in grain yield with the combined application of FYM and inorganic fertilizers were reported in other studies on sorghum (Hegde, 1996), rice (Sharma and Mittra, 1991), soybean (Reddy et al., 1999), maize (Ma et al., 1999), and wheat (Yadav et al., 2000; Yadav, 2001).

Table 6: Effect of the combined use of FYM and inorganic fertilizers on sorghum grain yield (kg ha^{-1}) at Kobo, pooled over five years

	Inorganic fertilizer (N/P kg ha^{-1})			
FYM (t ha^{-1})	0/0	21/10	41/20	Mean
0	3340g	3362fg	3726de	3476
5	3549efg	3822cd	4075b	3816
10	3457fg	3997bc	4067b	3841
15	3590def	4194b	4535a	4106
Mean	3484	3844	4101	

Means within columns and rows followed by the same letters are not significantly different at $p \leq 0.05$.
Source: Bayu et al. (2006b)

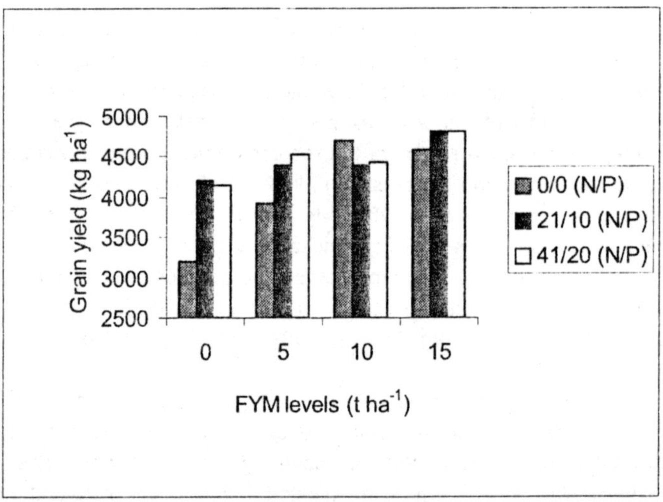

Figure 1: Effect of FYM and inorganic fertilizers on the grain yield of sorghum at Chefa, combined over years.
Source: Alemu & Bayu (2005)

Table 7: Effect of integrated application of FYM and inorganic N fertilizer on the grain yield of sorghum and maize at Sirinka

Mineral N fertilizer (kg N ha^{-1})	FYM levels (t ha^{-1})				
	0	10	20	30	Mean
			Maize		
0	1145i	2010h	2337gh	3613de	2276
40	1397i	2742fg	3100ef	3751de	2748
80	3215ef	3686de	4927ab	5230a	4264
120	944cd	4371bc	5082a	5249a	4662
Mean	2426	3202	3862	4461	
			Sorghum		
0	1619	2116	2326	2667	2182b
40	2041	2439	2861	3054	2599a
80	2332	2783	2733	2944	2698a
120	2468	2847	2428	2589	2583a
Mean	2115c	2546b	2587ab	2814a	

Means followed by the same letters are not significantly different at $p \leq 0.05$.
Source: Bayu et al. (2004)

4.4.2 Stover yield

Like grain yield several research results have shown increase in stover yield with FYM application. For instance, Bayu et al. (2006b) reported significant increase in stover yield due to FYM application at Kobo, where the application of 10 and 15 t FYM ha^{-1} produced significantly greater stover yield compared to the FYM control (Table 8). Alemu & Bayu (2005) similarly reported significant ($p \leq 0.01$) increase in stover yield at Chefa with the combined application of FYM and inorganic fertilizer (Figure 2). At Chefa, as high as 11 to 42% increase in stover yields were obtained with the application of 5, 10 and 15 t FYM ha^{-1} along with 41/20 kg N/P ha^{-1}. Again application of 10 t FYM ha^{-1} resembled mineral fertilizer equivalences of more than 41 kg N ha^{-1} and 20 kg P ha^{-1}, respectively.

Table 8: Effect of the combined use of FYM and inorganic fertilizers on sorghum stover yield (kg ha^{-1}) at Kobo, pooled over five years

FYM (t ha^{-1})	Inorganic fertilizer (N/P kg ha^{-1})			
	0/0	21/10	41/20	Mean
0	3295a	4101a	4162a	3853C
5	3881a	4003a	4631a	4172BC
10	3966a	4422a	4435a	4274AB
15	4029a	4725a	5189a	4648A
Mean	3793B	4313A	4605A	

Means followed by the same letter are not significantly different at $p \leq 0.05$.
Source: Bayu et al. (2006b)

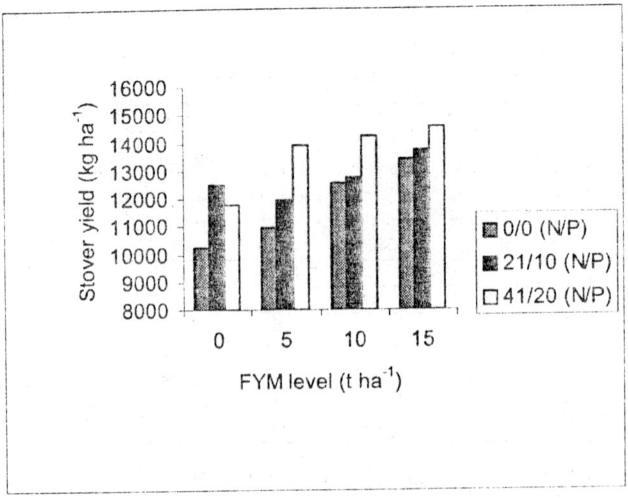

Figure 2: Effect of FYM and inorganic fertilizer on the stover yield of sorghum at Chefa, combined over years.
Source: Alemu & Bayu (2005)

4.5 Nitrogen use

4.5.1 Nitrogen Uptake

FYM application often leads to increase in nitrogen uptake due to the build-up of organic N in the soil from repeated applications, slow N release throughout the growing season, and from the development of a higher rooting density associated with the improved physical conditions of the soil, which in turn might enhance the nutrient-absorption capacity of the crop (Yadav, 2001). In a study conducted at Kobo significant N uptake increase as high as 29% in the grain and 57% in the stover of sorghum were recorded by applying FYM at 5, 10, and 15 t ha^{-1} (Table 9). Similarly, Xie and Mackenzie (1986) and Schlegel (1992) reported increased N uptake with FYM application.

Table 9: Effect of the combined use of FYM and inorganic fertilizers on nitrogen uptake at Kobo

FYM (t ha^{-1})	*Grain N uptake (kg ha^{-1})*	*Stover N uptake (kg ha^{-1})*
0	*76.1b*	*25.1c*
5	*91.4a*	*30.9b*
10	*91.3a*	*35.9ab*
15	*98.0a*	*39.4a*

Means followed by the same letter are not significantly different at $p \leq 0.05$.
Source: Adapted from Bayu et al. (2006b)

4.5.2 Nitrogen Use Efficiency

The combined use of FYM with inorganic fertilizers improves the efficient use of the inorganic fertilizer by the crop. The improved efficiency of inorganic fertilizer N when used in conjunction with FYM could be due to the fact that high rooting density due to improved physical conditions of the soil and greater water availability might enhance nutrient absorption capacity of the crop, thereby improving biological yield at a given level of fertilizer application (Boparai et al., 1992). Increased nitrogen use efficiency ranging between 21 and

73 kg kg^{-1} was obtained at Kobo when 21/10 kg ha^{-1} N/P fertilizer was applied together with 5, 10 and 15 t FYM ha^{-1} compared with 1.4 kg kg^{-1} when 21/10 kg ha^{-1} N/P fertilizer was applied alone (Table 10). Similarly, the combined application of 41/20 kg ha^{-1} N/P fertilizer together with 5, 10 and 15 t FYM ha^{-1} increased nitrogen use efficiency in the range of 14-38 kg kg^{-1} compared with 8 kg kg^{-1} when 41/20 kg ha^{-1} N/P fertilizer was applied alone. Yamoah et al. (2002) also reported increased fertilizer use efficiency when crop residues and inorganic fertilizers were combined than when fertilizer was applied alone. These results demonstrate that greater yields are attainable with a limited amount of inorganic fertilizer, if FYM is integrated in the nutrient supply system.

Table 10: Effect of the combined use of FYM and inorganic fertilizer on nitrogen use efficiency (kg kg^{-1})

FYM (t ha^{-1})	*21/10 kg ha^{-1} N/P*	*41/20 kg ha^{-1} N/P*
0	*1.4d*	*8.4d*
5	*21.1cd*	*16.3cd*
10	*73.4a*	*13.7cd*
15	*57.2ab*	*37.5bc*

Means followed by the same letter are not significantly different at $p \leq 0.05$.
Source: Adapted from Bayu et al. (2006b)

4.6 Weeds

4.6.1 Striga infestation

Striga infestation is usually heavier in nutrient poor soils that have been exhausted by several years of cultivation, soil erosion and poor input uses (Vogt et al., 1991). In north eastern Ethiopia poor soil nutrient status have already led to the build-up of the Striga seed bank in the soil. Thus, the increasing throat of Striga in the area could partly be attributed to the decline in soil fertility. In spite of the conflicting reports from different workers, the majority seem to suggest that applied nitrogen and farmyard manure increase crop vigor and directly or indirectly improve grain yields under Striga attack (Agbobli, 1991; Osman et al., 1991).

Results of a study conducted to see the effect of the combined use of FYM and nitrogen fertilizer on striga infestation showed that FYM application at 20 and 30 t ha^{-1} in maize have significantly reduced striga infestation by up to 70% compared with the control (Table 11). The same study showed that in sorghum the combined application of 40 and 80 kg N ha^{-1} each with 30 t FYM ha^{-1} and 120 kg N ha^{-1} with 10, 20 and 30 t FYM ha^{-1} resulted in significantly reduced striga infestation (Table 12).

Table 11: Effects of inorganic N fertilizer and FYM on striga infestation in maize combined over three years

N levels (kg ha^{-1})	Shoot 18.75 m^{-2}	FYM levels (t ha^{-1})	Shoot 18.75 m^{-2}
0	59a	0	63a*
40	48ab	10	46ab
80	29b	20	32bc
120	25b	30	19c

*Means followed by the same letters are not significantly different at $p \leq 0.05$.
Source: Bayu et al. (2004)

Table 12: Effects of inorganic N fertilizer and FYM on striga infestation (shoots 18.75 m^{-2}) in sorghum combined over two years

N levels (kg ha^{-1})	FYM levels (t ha^{-1})				
	0	10	20	30	Mean
0	58.0b	37.3bc	40.0bc	11.5c	37
40	124.7a	37.3bc	26.0bc	12.7c	50
80	21.2bc	58.7b	18.8bc	8.3c	27
120	16.0bc	6.7c	12.8c	8.7c	11
Mean	55	35	24	10	

Means followed by the same letters are not significantly different at $p \leq 0.05$.
Source: Bayu et al. (2004)

4.7 Concluding Remarks

Application of FYM improved several soil physicochemical properties. Total N, available P, exchangeable K and organic carbon contents of the soil and soil water content were substantially improved. The build-up of soil nutrients through repeated application of FYM was quite considerable for an area with a hot tropical climate like north eastern Ethiopia where organic matter decomposition and loss were expected to be high. The results revealed that application of FYM could improve the physicochemical properties of the soil within a reasonably short period of time. Thus, adding FYM appears to be an attractive strategy for resource-poor farmers to sustain soil productivity. Furthermore, the combined use of FYM and inorganic fertilizer increased crop yield at reduced inorganic fertilizer input while maintaining the soil resource. The combined application of 80 and 120 kg N ha^{-1} with 20 and 30 t FYM ha^{-1} also significantly reduced Striga infestation. The availability of such relatively high amount of FYM is sometimes questionable. However, one reality that should be noted is that although the rate of manure applications, which is usually recommended, on hectarage basis is large small scale farmers in Africa may actually need a small amount as they cultivate very small areas. Besides, as a scarce resource, having multiple uses, farmers may need to apply manure to economically important crops. It is also a type of input that may need to be applied every two or three years, rather than applying every year.

References

Agbenin, J.O. & Goladi, J.T. (1997) Carbon, nitrogen and phosphorus dynamics under continuous cultivation as influenced by farmyard manure and inorganic fertilizers in the savanna of northern Nigeria. Agric. Ecosys. Environ. **63**, 17-24

Agbobli, C.A. (1991) Effect of nitrogen rates on *Striga asiatica* emergence in maize culture in Togo. Ransom, J.K.; Musselman, L.J.; Worsham, A.D. & Parker, C (eds.) Proceedings of the 5th International Symposium on Parasitic Weeds. Nairobi, Kenya, 28-30

Alemu, G. & Bayu, W. (2005) Effects of farmyard manure and combined N and P fertilizers on sorghum and soil characteristics in north eastern Ethiopia. J. Sust. Agric. **26(2)**, 23-41

Bayu, W.; Reda, F.; Esilaba, A.O. & Ransom, J.K. (2004) Observations on the control of Striga through integrated nutrient management in Welo. Arem **VI**, 49-57

Bayu, W.; Rethman, N.F.G.; Hammes, P.S. & Alemu, G. (2006a) Application of farmyard manure improved the chemical and physical properties of the soil in a semi-arid area in Ethiopia. Biol. Agric. Hort. **24**, 293-300

Bayu, W.; Rethman, N.F.G.; Hammes, P.S. & Alemu, G. (2006b) Effects of farmyard manure and inorganic fertilizers on sorghum growth, yield, and nitrogen use in a semi-arid area of Ethiopia. J. Plant Nutr. **29**, 391-407

Belay, A.; Claassens, A.S.; Wehner, F.C. & De Beer, J. M. (2001) Influence of residual manure on selected nutrient elements and microbial composition of soil under long-term crop rotation. S. Afr. J. Plant Soil **18**, 1-6

Boparai, B.S.; Singh, Y. & Sharma, B.D. (1992) Effect of green manuring with Sesbania aculeate on physical properties of soil and growth of wheat in rice-wheat and maize-wheat cropping sequence in semi-arid regions of India. Arid Soil Res. Rehab. **6**, 135-143

Carter, D.C.; Harris, D.; Yongquist, J.B. & Persaud, N. (1992) Soil properties, crop water use and cereal yields in Botswana after additions of mulch and manure. Field Crops Res. **30**, 97-109

Esilaba, A.O.; Reda, F.; Ransom, J.K.; Bayu, W.; Woldewahid, G. & Zemichael, B. (2000) Integrated nutrient management strategies for soil fertility improvement and striga control in northern Ethiopia. Afr. Crop Sci. J. **8(4)**, 403-410

Gill, H.S. & Meelu, O.P. (1982) Studies on the substitution of inorganic fertilizers in rice-wheat rotation. Fert. Res. **3**, 303-314

Gomez, K. A. & Gomez, A. A. (1984) Statistical procedures for agricultural research. New York: John Wiley and Sons

Goyal, S.; Mishra, M.M.; Hooda, I.S. & Singh, R. (1992) Organic matter-microbial biomass relationship in field experiments under tropical conditions: Effects of inorganic fertilization and organic amendments. Soil Biol. and Bioch. **24,** 1081-1084

Hegde, D.M. (1996) Long-term sustainability of productivity in an irrigated sorghum-wheat system through integrated nutrient supply. Field Crops Res. **48**,167-175

Hoffmann, I.; Gerling, D.; Kyiogwom, U.B. & Mane-Bielfeldt, A. (2001) Farmers' management strategies to maintain soil fertility in a remote area in northwest Nigeria. Agric. Ecosyst. Environ **86**, 263-275

Kaihura, F.B.S.; Kullaya, I.K.; Kilasara, M.; Aune, J.B.; Singh, B.R.; Lal, R. & Arshad, M.A. (1999) Soil quality effects of accelerated erosion and management systems in three ecoregions of Tanzania. Soil Till. Res. **53**, 59-70

Lungu, O. I.; Temba, J.; Chirma, B. & Lungu, C. (1993) Effects of lime and farmyard manure on soil acidity and maize growth on an acid Alfisol from Zambia. Tropical Agric. (Trinidad) **70**, 309–314

Lupwayi, N.Z. & Haque, I. (1999) Leucaena hedgerow intercropping and cattle manure application in the Ethiopian highlands. III. Nutrient balances. Biol. Fert. Soils **28**, 204-211

Ma, B.L.; Dwyer, L.M. & Gregorich, G.E. (1999) Soil nitrogen amendment effects on nitrogen uptake and grain yield of maize. Agron. J. **91**, 650-656

Mokwunye, A.U. (1991) Alleviating soil fertility constraints to increased crop production in West Africa. Kluwer Axcademic Publisher, Dordrecht, The Netherlands

Osman, M.A.; Raju, P.S. & Peacock, J.M. (1991) The effect of soil temperature, moisture and nitrogen on *Striga asiatica* seed germination, viability and emergence on sorghum (*Sorghum bicolor*) roots under field conditions. Plant and Soil **131**, 265-273

Page, A.L.; Miller, R.H. & Keeney, D.R. (1982) Methods of soil analysis. Part 2. Chemical and microbiological properties, 2nd edition. Madison, WI: Soil Science Society of America

Powell, J.M. (1986) Manure for cropping: a case study from central Nigeria. Expl. Agric. **22**, 15-24

Prasad, P.V.V.; Satyanarayana, V.; Murphy, V.R.K. & Boote, K.J. (2002) Maximizing yields in rice-groundnut cropping sequence through integrated nutrient management. Field Crops Res. **75**, 9–21

Reddy, D.D.; Rao, A.S.; Reddy, K. & Takkar, P.N. (1999) Yield sustainability and phosphorus utilization in soybean-wheat system on vertisols in response to integrated use of manure and fertilizer phosphorus. Field Crops Res. **62**, 181-190

Satyanarayana, V.; Prasad, P.V.V.; Murthy, V.R.K. & Boote, K.J. (2002) Influence of integrated use of farmyard manure and inorganic fertilizers on yield and yield components of irrigated lowland rice. J. Plant Nutri. **25**, 2081–2090

Schlegel, A.J. (1992) Effect of composted manure on soil chemical properties and nitrogen use by grain sorghum. J. Prod. Agric. **5**, 153-157

Schjonning, P.; Christensen, B.T. & Carstensen, B. (1994) Physical and chemical properties of a sandy loam receiving animal manure, mineral fertilizer or no fertilizer for 90 years. Eurp. J. Soil Sci. **45**, 257-268

Sharma, A. R. & Mittra, B.N. (1988) Effect of combinations of organic materials and nitrogen fertilizer on growth, yield and nitrogen uptake of rice. J. Agric. Sci. (Cambridge) **111**, 495–501

Sharma, A.R. & Mittra, B.N. (1991) Effect of different rates of application of organic and nitrogen fertilizers in a rice-based cropping system. J. Agric. Sci. (Cambridge) **117**, 313–318

Shirani, H.; Hajabbasi, M.A.; Afyuni, M. & Hemmat, A. (2002) Effects of farmyard manure and tillage systems on soil physical properties and corn yield in central Iran. Soil Tillage Res. **68**, 101-108

Vogt, W.; Sauerborn, J. & Honisch, M. (1991) *Striga hermonthica* distribution and infestation in Ghana and Togo on grain crops. Ransom, J.K., Musselman, L.J., Worsham, A.D. & Parker, C (eds.) Proceedings of the 5th International Symposium on Parasitic Weeds. Nairobi, Kenya, 372-377

Walkley, A. & Black, I. (1934) An examination of the Degtejareff method for determining soil organic matter and a proposed modification of the chromic acid titration method. Soil Sci. **37**, 29-38

Warman, P.R. & Cooper, J.M. (2000) Fertilization of a mixed forage crop with fresh and composted chicken manure and NPK fertilizer: Effects on soil and tissue Ca, Mg, S, B, Cu, Mn and Zn. Can. J. Soil Sci. **80**, 345-352

Whalen, J.K.; Chang, C. & Clayton, G.W. (2002) Cattle manure and lime amendments to improve crop production of acid soils in northern Alberta. Can. J. Soil Sci. **82**, 227-238

Wong, M.T.F.; Nortcliff, S. & Swift, R.S. (1998) Method of determining the acid ameliorating capacity of plant residue compost, urban waste compost, farmyard manure, and peat applied to tropical soils. Commun. Soil Sci. Plant Anal. **29**, 2927-2937

Xie, R. & Mackenzie, A.F. (1986) Urea and manure effects on soil nitrogen and corn dry matter yields. Soil Sci. Soc. Am. J. **50**, 1504-1508

Yadav, R.L. (2001) On-farm experiments on integrated nutrient management in rice-wheat cropping systems. Expl. Agric. **37**, 99-113

Yadav, R.L.; Dwivedi, B.S.; Prasad, K.; Tomar, O.K.; Shurpali, N.J. & Pandey, P.S. (2000) Yield trends, and changes in soil organic-C and available NPK in a long-term rice-wheat system under integrated use of manures and fertilizers. Field Crops Res. **68,** 219-246

Yamoah, C.F.; Bationo, A.; Shapiro, B. & Koala, S. (2002). Trend and stability analysis of millet yields treated with fertilizer and crop residues in the Sahel. Field Crops Res. **75**, 53-62

Zia, M.S.; Munsif, M.; Aslam, M. & Gill, M.A. (1992) Integrated use of organic manures and inorganic fertilizers for the cultivation of lowland rice in Pakistan. Soil Sci. Plant Nutr. **38**, 331-338

Optimising cropping systems with green manure and tillage techniques in Organic Farming in Kenya

Rhoda Jerop Birech[1]; *Bernhard Freyer*[2]

University of Natural Resources and Applied Life Sciences, BOKU. Gregor Mendel Strasse, 1180 Vienna, Austria, www.nas.boku.ac.at/oekoland.html
rhodajerop@yahoo.com[1]; *Bernhard.Freyer@boku.ac.at*[2] *(corresponding author)*

5.0 Summary

Food security and ecological stability in the tropics can be assured when sustainable soil fertility management approaches are used. Inputs derived from locally available organic sources have a profound potential to provide plant nutrients and to improve the physical status and the biological activity of a soil. A two-year experiment was set up at Egerton University, situated in the highlands of Central Rift Valley to study the effect of different green manures and their application methods on the yield of organically grown wheat. The effect of green manures obtained from *Lablab* (*Dolichos lablab*), *Tithonia* (*Tithonia diversifolia*) and *Tephrosia* (*Tephrosia vogelii*) on the yield of organic wheat was evaluated. The average yield of wheat after applying *Lablab* was 4064 kg/ha, *Tithonia* gave 4041 kg/ha, *Tephrosia* gave 3717 kg/ha, while the control yielded 3549 kg/ha. Furthermore, wheat grain yield increased by 1000 kg ha^{-1} after mulching of the green manures compared to when they were incorporated into the soil to a depth of 30 cm. The implications of the use of green manures in different cropping systems in the tropics are discussed. Formulation of ecologically sound and economically viable soil and crop management options could enhance the adoption of Organic Farming systems in Kenya. The investigation underlines how important it is to use best knowledge for optimal Organic Farming. Furthermore, the results have shown that Organic Farming is able to produce yields which are comparable with conventional farming results.

Keywords: mulch, legumes, biomass transfer, wheat

5.1 Introduction

5.1.1 Problem statement as per the situation in the tropics

Food production is low and declining in tropical climates and particularly sub-Saharan Africa. This has been attributed to population increase which has put pressure on scarce land's resources, particularly soils, forests and water. Soil erosion, continuous cultivation without due replenishment of nutrient stocks, limited nutrient inputs, lack of high quality organic manures and plant residues have collectively led to nutrient mining and a decline of soil fertility (Jama et al., 2000). In traditional agriculture, arable land could be left fallow for as long as 30 years to allow soil to accumulate nutrients and to acquire self-rejuvenation. Such long duration fallows have disappeared in most parts of the tropics because of competing land use demands. Shorter duration fallow periods are not sufficient to restore the soil nutrient pools and soil organic matter levels sufficient to support economic crop yields. The conventional approach to improving soil fertility by adding synthetic fertilizers has not been sustainable because it does not address soil fertility in its totality; i.e. chemical, physical and biological fertility; it often exhausts the humus content and makes the soil prone to soil erosion. Other key reasons typical of small scale agriculture that limit the use of external input are the high costs involved and the negative marginal returns received upon use. Moreover, the inorganic fertilizers are often too expensive for the subsistence farmers representing the majority of the farming population in developing countries. There is therefore need to integrate sustainable soil management practices that focus more strongly on the use of locally available resources such as green manures, crop residues and animal manures rather than expensive external inputs.

Nitrogen (N) is highly mobile in tropical climates and can easily be lost by leaching and volatilization. This loss is reported to be higher in mineral fertilizer-derived N than in organically-derived N (Vanlauwe et al., 2001). Additionally, N is taken up in highest quantities by plants compared to most other plant nutrients. These properties make N deficiency a common phenomenon in tropical soils and it is therefore a major factor responsible for low productivity in these regions. The average annual mining of N in parts of western Kenya is up to 112 kg N ha^{-1} $year^{-1}$ (Smaling & Braun, 1996). Legumes therefore have a great potential to supply most, if not all of the N requirements of crops in a cropping system owing to their ability to fix atmospheric N (Niang et al., 2002). When adopted under organic management, legumes can secure crop yields, which are the basis for a family nutrition and income.

This paper reports a two-year experiment done at Egerton, Central Rift Valley, Kenya, to identify green manure species and mode of application for sustainable Organic Farming. The implications of the use of these materials in different cropping systems that are common in the tropics are discussed.

5.1.2 The organic cropping system approach

5.1.2.1 Building a stable soil system

Emerging scenario of declining soil productivity and food deficit in parts of the tropics demand the identification of food production systems that enrich the soils, that optimize ecological balance and that are economically viable to resource-poor farmers. The use of appropriate organic inputs is an option to providing plant nutrients as well as building long-term soil fertility and ascertaining environmental integrity. Products of organic matter decomposition add plant nutrients and carbon, stabilise the soil structure and regulate soil pests. They are substrates for soil organisms and thus contribute to ecosystem stability. To optimize sustainable soil fertility within the realities of the current farming systems, Organic Farming advocated for a soil fertility management packages that emulate the natural systems with reduced external inputs, especially synthetic fertilizers which might pollute the environment. The use of soft plant tissues such as leaves, twigs and petioles, usually referred to as green manures, is common in agriculture owing to their ease of decomposing and releasing embedded nutrients. The extent to which green manures have an impact on soil productivity depends on the cropping system, material quality, the climate and the application method.

5.1.2.2 Green manure approach

Green manures can be annual or perennial herbaceous plants grown in a pure or mixed stand during the whole or part of the year. Of the entire plant kingdom, legumes, whether grown for grain, pasture or as trees in an agro-forestry system, have the potential to add nitrogen to a soil-plant-system (Prata Neves et al., see this volume). Apart from legumes, other plants like *Tithonia diversifolia* (Cong & Merckx, 2005) have been shown to have the ability to source for plant nutrients from the soils and to release them readily in plant-available forms whenever it is applied as a green manure. Plants with deep roots also have the ability of capturing leached nutrients beyond the reach of shallow rooted crop and the use of these materials as green manure enrich crops (Graves et al., 2004).

The success of green manures in soil fertility management depend on their adaptability to the local climates, flexibility to fit into the existing cropping system such as rotation or intercropping with major food crops, N_2-fixing potential and net nutrient input into the soil. Another major criterion which these plants must meet in order to enhance adoption by farmers includes their ability to provide certain goods in terms of grain and/or fodder in addition to supporting soil fertility (Ojiem et al., 2007). This underscores the need to screen legumes and other green manure plants for their potential to fit into the dictates of the climatic bounds as well as socio-economic and cultural environments. Most of the green manure evaluations in Kenya have focused on their potential as a component of integrated nutrient management (Mureithi et al., 2003), where they are used in combination with fertilizers but not of Organic Farming systems where these organic resources are the principle sources of N nutrition. There is therefore need to identify appropriate green manure species and residue management techniques that can recycle sufficient plant nutrients which may be used to formulate simple recommendation packages for farmers who wish to convert to Organic Farming. The availability of such management techniques will be a big step to support the young organic sector in Kenya.

5.1.2.3 Production of green manures

The crop-green manure production interface can take different dimensions. Parallel production of both the food/cash crop and the green manure may be done on different pieces of land at the same time; the manures are then later brought into the crop-land through cut-and-carry method or biomass transfer. Production may also take place simultaneously or sequentially. In the case of simultaneous production, both the food/cash crop and the green manure crop are intercropped together in the same piece of land at the same time. Sequential cropping is done when crops are planted in rotation with green manures a system that is referred to as fallowing.

A short duration fallow takes 3 months to 12 months and involves the use of fast-growing plant species, which can accumulate substantial biomass within a stipulated time. An experiment in western Kenya using mucuna (*Mucuna purpurea*) and crotalaria (*Crotalaria ochroleuca*) showed the legumes to accumulate about 135 kg/ha N in their shoot biomass within 3-4 months of growth. This suggests that these legumes under short duration fallow system have the potential to supply most of the requirements of subsequent cereal crops (Ojiem et al., 2007). On the contrary, long-duration shrubby species, which take three or more seasons to accumulate sufficient biomass are better utilized when grown outside the crop land either along the hedges or in separate land set aside for this purpose. Success has been reported with the use of biomass from shrubs such as *Tithonia diversifolia*, *Gliricidia sepium*, *Calliandra calothyrsus*, and *Leucaena leucocephala* (Graves et al., 2004). For biomass transfer techniques to be adopted by farmers, there has to be sufficient land aside of the crop land available to produce the required biomass. This is likely to be a limitation if land is scarce and farms are small.

5.1.2.4 Biomass quality

The rate at which nutrients are released from green manures after application influence the responses of subsequent crops. Green manures whose nutrients are released within the first season after application are considered to be of high quality while the reverse is true. The polyphenol, lignin and N contents of biomass control N mineralization and hence they are indicators of biomass quality (Palm et al., 2001). The critical levels of polyphenol, lignin and N concentration required for optimal N mineralization are <4%, <15% and >2.5% respectively. The concentration of quality indicators depends on the species and the developmental stage of the biomass being applied. Younger plant tissues have less carbon and lignin deposits than older plant parts. Conversely, quick release of nutrients also poses a possibility for N loss by leaching and low residual effect in subsequent years.

Vanlauwe et al., (2001) provide a consolidated perspective on biomass quality based on experiments that were performed in East, West and southern Africa. In these trials, organic materials with varying N concentrations (0.14 – 3.53%) were applied on maize using biomass transfer method (materials brought from outside the farm). The trials showed a positive relationship between N concentration and N uptake by the first generation of maize. Low quality material such as maize stover immobilized N and performed less than the control during this period. However, N uptake by the second year maize was higher from the hitherto low quality material. Thus, high quality organic materials may stimulate plant growth within the first season of application, but medium to low quality materials release their N slowly and have greater residual effects. Nevertheless, farmers have been shown to reject material whose nutrients cannot be released within the first season after incorporation (Vanlauwe et al., 2001) suggesting a need for other management strategies such as composting for low quality materials.

5.1.2.5 Application method

Not all nutrients embedded in plant tissues are available for plant uptake soon after they are applied. N recoveries from legume residues during the first season can be as low as 10-20% of the total tissue N (Ojiem et al., 2007). To enhance residue decomposition and nutrient release from residues, certain management strategies, such as reduction of particle size help to increase the surface area for microbial action. Incorporation of material into the soil is also perceived to increase the decomposition rate since the material is placed in contact with favourable moisture and temperature.

5.2 Materials and Methods

5.2.1 Site description

To study the management of organic materials under biomass transfer cropping system at a sub-humid tropical environment, a two-year field trial was established at Egerton University, in the highlands of the Central Rift Valley, Kenya (0^0 13'S, $35^0$30'E; 2200 m above sea level). The site has an average annual temperature of 17.5^{0}C and annual rainfall of 1017 mm. The region experiences bimodal type of rainfall with two rainfall peaks; the long and the short rains. The amount of rainfall received during the growth period (April – September) was 591 mm and 455 mm in 2005 and 2006, respectively. Lima bean was planted after harvesting wheat from October 2005-April 2006 in the short rains, but there was a total crop failure because the rain was too little with 164mm.

5.2.2 Experimental design and treatments

The trial started in April 2005. Green manure (leaves and soft twigs) of *Tithonia* (*Tithonia diversifolia*), *Tephrosia* (*Tephrosia vogelii*) and *Lablab* (*Dolichos Lablab*) obtained from outside the farm were applied at a uniform rate of 2 tons ha^{-1} on a dry matter basis. An untreated control was included. The plant material was chopped into small pieces (2 -4 cm) and applied using three different treatments: mulching and incorporation into the soil to a depth of 15 cm and 30 cm, respectively. The experiment was laid out in a Randomized Complete Block Design (RCBD) with 4 replications and repeated for two years. Two weeks after application of the plant materials, rock phosphate at a rate of 20 kg P ha^{-1} was applied and wheat was sown at an inter-row spacing of 20 cm with a sowing depth of about 3 cm. Wheat grains were harvested at the end of September and grain yield (13% moisture content) was recorded. Above ground dry straw biomass was determined after harvesting grains. Soil available N was determined at grain-filling stage using the steam distillation and titration procedure (Keeney & Nelson, 1982). All data were analysed statistically using SPSS (version 11.5) and means were separated by the Tukey test.

5.3 Results

5.3.1 Effect of different green manures on wheat

In the first year *Tithonia* resulted in higher wheat grain yield (4764 kg/ha), being 18% higher than the control (4032 kg/ha). In the following year, *Lablab* application led to the highest wheat grain yield (3684 kg/ha) that was 20% higher compared with the control (3065 kg/ha) (Table 1).

Table 1: Wheat dry matter and grain yield after green manure application (all application methods)

	2005		2006	
Shoot biomass	Average wheat grain yield kg ha $^{-1}$	Dry straw biomass kg ha $^{-1}$	Average wheat grain yield kg ha $^{-1}$	Soil available N at grain-filling mg kg $^{-1}$
Dolichos lablab	4444a	4634a	3684a	78a
Tithonia diversifolia	4764a	4610a	3318a	89a
Tephrosia vogelii	4504a	4610a	2930a	91a
Control	4032a	4500a	3065a	70a
Mean	4436	4589	3249	82
Standard deviation	589	749	968	26

Means in a column followed by the same letter are not significantly different at P<0.05

However, grain yields of wheat were not significantly different. The average yield for all treatments was lower for the second year compared with the first year, a factor that may be attributed to lower rainfall in the second year. Taking the averages of the two years into account, the wheat grain amounted to 4064 kg/ha, 4041 kg/ha and 3717 kg/ha, respectively after applying *Lablab*, *Tithonia* precrop and *Tephrosia residues*. The control resulted in 3549 kg/ha grain yield of wheat. Straw dry matter yield was not significantly different across the three green manure sources. Available N concentration in the top soil taken at grain-filling was lowest in the control plots (70 mg N/kg soil) but this effect was not statistically significant as well.

5.3.2 Effect of biomass incorporation depth on wheat

Mulching of green manures resulted in higher wheat grain yields as well when compared with the incorporation at 30 cm soil depth and the control in both years. During the first year, mulching and shallow incorporation of green manures seemed to be advantageous when compared with deeper incorporation of residues to 30 cm soil depth, although it was not statistically significant. The mulch treatment gave also significantly higher (P<0.05) wheat straw yield than incorporation at 30 cm and the control in the first year. During the second year, mulching and shallow incorporation resulted in significantly higher (P<0.05) wheat grain yields than deep incorporation and the control. Wheat grain yield was about 1000 kg/ha higher after mulching than after incorporation at 30 cm (Table 2).

Table 2: Effect of biomass management on yield and yield components of wheat (average of all applications)

	2005		2006	
Incorporation depth cm	Average wheat yield kg ha^{-1}	Dry straw biomass kg ha^{-1}	Average wheat yield kg ha^{-1}	Available N at grain-filling mg kg^{-1}
Incorporated at 15 cm	4589a	4494ab	3222ab	94a
Incorporated at 30 cm	4493a	4309a	2857a	81a
Surface mulch	4630a	5051b	3853b	84a
Control	4032a	4500a	3065a	70a
Mean	4436	4589	3249	82
Standard deviation	589	827	968	26

Means in a column followed by the same letter are not significantly different at $P<0.05$

5.4 Discussion

5.4.1 Green manure effect

Different green manures have different nutrient concentrations and variable rates of nutrient mineralization. In this experiment, *Tithonia* and *Lablab* increased wheat grain yields when compared with *Tephrosia* and the untreated control. On the overall, the three biomass sources resulted in better yields than the control. Since the site had been under natural fallow for one year prior to the trial, the lack of significances among the treatments may be a result of a relatively fertile soil. Research reports on green manures in Kenya showed little treatment effect in areas with high inherent soil fertility (Mureithi, 2003).

Lablab is one of the few legumes among those studied in Kenya with many uses. Many trials in the tropics depict the multiple use of *Lablab*, which qualifies it as a grain legume (Ojiem, 2007), green manure (Mureithi, 2003), and fodder legume (Hardason & Atkins, 2002). Furthermore, it is adapted to most climates in Kenya except the dry areas. In a legume research network in Kenya, *Lablab* was found to be adapted to most sites in Central Highland, South West Highlands, Western Region, Coastal Lowland and the Lake Victoria basin, all with moderate to high rainfall levels, except is the semi-arid eastern Kenya where pigeon pea (*Cajanus cajan*) is better adapted (Mureithi, 2003). An experiment conducted at Egerton showed *Lablab* residue incorporation to significantly increase the grain yield of subsequent maize compared to the incorporation of natural fallow, garden peas (*Pisum sativum*), and crotalaria (*Crotalaria ochroleuca*) residues (Lelei, 2004). In this trial, *Lablab* nutrient contents in shoots were: 32 g kg^{-1} N, 21 g kg^{-1} P and 15.7 g kg^{-1} K (Lelei, 2004). Similar results were obtained in Rongai, Kenya, when *Lablab* proved better than chick peas (*Cicer arietinum*), field beans (*Phaseolus vulgaris*), soybean (*Glycine max*) and field peas (*Pisum sativum*) in influencing the yield of subsequent maize (Cheruiyot *et al.*, 2003).

The advantage of *Tithonia* may be attributed to higher nutrient concentration in tissues and their quick release of the same into plant-available forms. Trials in Maseno district of western Kenya revealed *Tithonia* leaves to have 30 g kg^{-1} nitrogen (N), 1.8 g kg^{-1} phosphorus (P), 46 g kg^{-1} potassium (K), 19.1 g kg^{-1} calcium (Ca) and 3.7 g kg^{-1} magnesium (Mg). The shrub had high concentrations of P, K and Ca compared to *Tephrosia* whose leaves had 30.5 g kg^{-1} N, 1.2 g kg^{-1} P, 15.7 g kg^{-1} K, 13.0 g kg^{-1} Ca and 4.7 g kg^{-1} (Rutunga et al., 1999). Though *Tithonia* is a non-legume, the N content in its leaves is nearly the same as that of *Tephrosia*, a nitrogen fixing legume. The differences observed in the crop response may therefore be due to variation in nutrient concentrations in the green manures used.

5.4.2 Mode of application of organic materials

Management of N dynamics is essential in organic systems to limit N loss and to increase synchrony with crop demands. As a general rule, tillage hastens the breakdown of soil residues and raises the level of N availability (Hardason & Atkins, 2000), but in most of the parameters observed under the current study, the surface application of plant material (mulching) was more superior in influencing wheat performance. This was followed closely by the shallow incorporation at 15 cm in the case of wheat grain yields. This effect might be attributed to wheat root morphology. The lateral roots of wheat, which are responsible for absorption of water and nutrients, are concentrated within the upper 6 - 13 cm of the top soil (Kinyua et al., 2003). This implies that soil moisture and plant nutrients should be available at these shallow depths in order to maximize plant growth. The application of organic material as mulch or at a shallow depth is therefore likely to be the most appropriate green manure application method for shallow rooted crops like wheat especially in warm and humid environments. After all, the process of making of planting furrow and weeding somewhat mixed the mulch material with the top soil layer. Apart from making the nutrients inaccessible to feeder roots, deeper application at 30cm is likely to dilute the material with a larger soil mass while reducing their concentration in the soil layer of main uptake.

Other than root distribution, rainfall levels seem to play an important role on the choice of material application. In semi-arid tropics, mulching is important for conserving soil moisture. In an experiment done in a semi-arid area in eastern Kenya, mulching with legume biomass (sunhemp, mucuna and lima bean) gave higher maize yields (2.8t ha^{-1}) than incorporation (2.3t ha^{-1}) (not statistically significant), but in the central highlands with high rainfall, maize grain yield was highest after legume residue incorporation than mulching (Mureithi, 2003). In Kitale, Rift Valley, incorporation (5.1t ha^{-1}) resulted to nearly similar maize yields with mulching (4.8t ha^{-1}). Kaizzi & Wortmann (2001) also found nutrient release and decomposition of *Lantana camara, Senna hirsuta, Tithonia diversifolia* and *Aspilia kotschyi* to be faster in incorporation than surface application but found no crop yield differences. They concluded that plant materials similar to those studies were best when surface-applied because they reduce the cost of incorporation. Increasing interest in surface application rather than incorporation in sub-Saharan Africa has been reported (Schroth et al., 1992; Fischler & Wortmann, 1999).

5.4.3 Potential of green manures in Organic Farming systems in the tropics

There is an ample variety of plant species, mostly legumes that can be utilized for soil fertility improvement in the tropics and particularly in Organic Farming. The choice of species apparently depends on their niche adaptation to the local climate and farming system. However, species and variety selection in favour of high N_2-fixation and biomass production can be done on regional basis. At least *Tithonia* and *Lablab* green manures can support organic wheat production in semi-humid regions.

For green manures to support sustainable soil fertility management, the availability of these materials in sufficient quantities has to be assured. Short-duration green manure species which are adapted to the cooler and drier climate of Egerton (17^{0}C average temperature and

266 mm average rainfall during the short rains) could be produced on-farm during the short rain season. *Lablab* is among the few green manure legumes, which can accumulate about 6t ha^{-1} of shoot biomass at Egerton when it receives precipitation of about 400 mm within 5 months of its cycle. Rotating wheat with *Lablab* within a year-cycle is a feasible approach to soil fertility management in organic systems in Kenya. Within the climatic domains of the cooler tropics, the shrubby species like *Tithonia* and *Tephrosia*, which take a long time to accumulate sufficient biomass, may have to support soil fertility through transporting its biomass from outside the field. Research on cool-season green manures is needed.

Conventionally, legumes are known to form a component of integrated soil fertility management but not the sole source of plant nutrition especially in degraded soils. While they may fix atmospheric N sufficient to support the requirements of subsequent crops, P and K nutrition have to be supplied from outside the farm. Studies across many regions in Kenya showed that growing green manures required about 20 P kg ha^{-1} (Niang et al., 2002). P is essential for root and nodule formation. To meet this need, biomass transfer from species like *Tithonia*, being high in P and K and growing in unfarmed land like in farm hedges may recycle the nutrients to the farmland. Use of rock phosphate which is known to be deposited in four regions in East Africa is an important source of P for organic crop production. In fact, solubilization of P from rock P sources is enhanced by the action of organic acids present in organic manures.

Finally, the socio-economic valuation of biomass transfer approach showed that the high labour costs involved in transporting and applying biomass limit the adoption of the technology on a large scale and make the use of *Tithonia* on maize hardly profitable. However, use of these materials on high value crops such as vegetables and fruits (ICRAF, 1996) may spur adoption. The use of mulching instead of incorporation may also reduce production costs since soil operation like weeding achieves some degree of incorporation anyway. This, however, depends on crop species and specific climate.

5.5 Conclusion

This trial underlines how Organic Farming systems may be managed in semi-humid subtropical environments using biomass transfer and adapted mulch application of green manures. These approaches and techniques can easily be integrated in farms and are likely to support the organic production sector in Kenya. The trial as well as the discussion of other investigations has shown the importance of site-adapted plant selection. The empirical results led to three conclusions:

1. *Tithonia* and *Lablab* have the potential to support soil fertility for organic wheat production under the soil and climatic conditions in the Central highlands of the Rift valley, Kenya.
2. There is potential to screen plant species and their varieties for optimum biomass, nutrient yield and appropriate farming systems through which they may contribute to soil fertility management in the various climates of the tropics
3. The mode of application of green manures varies depending on green manure quality, climate and crop species.

References

Cheruiyot, E.K.; Mumera, L.M.; Nahone, L.N. & Mwonga, S.K. (2003) Effect of legume-managed fallow on weeds and soil nitrogen in following maize (*Zea mays L.*) and wheat (*Triticum aestivum L.*) crops in Rift Valley Highlands of Kenya. Australian Journal of Experimental Agriculture. **43** (6), 597-604

Cong, P.T. & Merckx, R. (2005) Improving P availability in two upland soils of Vietnam using *Tithonia diversifolia*. Plant and Soil **269**, 11–23

Fischler, M. & Wortmann, C.S. (1999) Green manures for maize-bean systems in Eastern Uganda: Agronomic performance and farmers' perceptions. Agroforestry Systems **47**, 123-138

Graves, A.; Matthews, R. & Waldie, K. (2004) Low external input technologies for livelihood improvement in subsistence agriculture. Advances in Agronomy **8**, 473–555

Hardason G. & Atkins, C. (2002) Optimizing biological N_2 fixation by legumes in farming systems. Plant and Soil **252**, 41-54

ICRAF (1996) International Centre for Research in Agroforestry: Annual Report. Nairobi

Jama, B.; Palm, C.A.; Buresh, R.J.; Niang, A.; Gachengo, C.; Nziguhbeba, G. & Amadalo, B. (2000) *Tithonia diversifolia* as a green manure for soil fertility improvement in Western Kenya: A review. Agroforestry Systems **49**, 201-221

Kaizzi, C.K. & Wortmann, C.S. (2001) Plant materials for soil fertility management in sub-humid tropical areas. Agronomy Journal of America **93**, 929-935

Keeney, D.R. & Nelson, D.W. (1982) Nitrogen inorganic forms. In: Methods of soil analysis, Part 2. Second edition. Page, A.L.; Miller, R.H. & Keeney, D.R. (Eds). 643-698. Agronomy Monogr. **9**. ASA and SSSA, Madison, WI, USA

Kinyua, M.G.; Njoka, E.M.; Birech, R.J. & Gesimba, R. (2002). Selection of drought tolerant bread wheat genotypes using root characteristics at seedling stage. International Journal of Agriculture and Rural Development **4**, 9-15

Lelei, J.N. (2004) Impact of soil amendments on maize performance and soil nutrient status in legume maize intercropping and rotation systems in Central Rift Valley Province of Kenya. Ph.D. thesis, University of Natural Resources and Applied Life Sciences, Austria

Mureithi, J.G.; Gachene, C.K.K. & Ojiem, J. (2003) The role of green manure legumes in small-holder farming systems in Kenya: The legume research network project. Tropical and Sub-tropical Agroecosystems **1**, 57-70

Niang, A.I.; Amadalo, B.A.; de Wolf, Y. & Gathumbi, S.M. (2002) Species screening for short-term planted fallows in the highlands of Western Kenya. Agroforestry Systems **56** , 145-154

Ojiem, J.O.; Vanlauwe, B.; Ridder, N & Giller, K.E. (2007) Niche-based assessment of contributions of legumes to the nitrogen economy of Western Kenya small-holder farms. Plant Soil **292**, 119-135

Palm., C.A.; Gachengo, C.N.; Delve, R.J.; Cadisch G. & Giller, K.E. (2001) Organic inputs for soil fertility management in tropical agroecosystems: application of an organic resource database. Agriculture, Ecosystems and Environment **83**, 27-42

Rutunga V.; Karanja. N.K.; Gachene C.K.K. & Palm C., (1999) Biomass production and nutrient accumulation by *Tephrosia vogelii* (Hemsley) A. Grey and *Tithonia diversifolia* Hook F. fallow during the six-month fallow period at Maseno, Western Kenya. Biotechnol. Agron. Soc. Environ. **3** (4), 237 – 246

Schroth, G.; Zech, W. & Heimann, G. (1992) Mulch decomposition under Agroforestry conditions in a sub-humid tropical savanna: processes and influence of perennial plants. Plant and Soil **147**, 1-11

Smaling, E.M.A. & Braun, A.R. (1996) Soil fertility research in sub-Saharan Africa: new dimensions, new challenges. Communications in Soil Science and Plant Analysis **27**, 365-386

Vanlauwe B.; Palm C.; Murwira H.K.; Merckx R. & Delve R. (2002) Organic resource management in sub-Saharan Africa: validation of a residue quality-driven decision support system. 17th WCSS conference held on August 2002 in Thailand

20 years of developing organic exports in the tropics; the need for and interface with research

Bo van Elzakker BSc Trop.Agr.
Director Tropics, Agro Eco, Postal address: P.O. Box 63, 6720 AB Bennekom, The Netherlands, E-Mail: b.vanelzakker@agroeco.nl

6.1 Introduction

The work in the tropics started when in the mid-1980's West European importers needed organic tropical products like coffee, cotton and bananas to supply the growing organic market. We were asked to assist those markets with getting product X certified organic. Originally, this was mainly in Latin-America (Mexico, Costa Rica, Argentina) and Asia (Philippines, Sri Lanka and India), but from the mid-nineties the work concentrated on sub-Saharan Africa. While some crops like oil palm, rubber and tea are grown on large estates, the focus in Africa shifted almost completely to smallholder agriculture. This chapter relates largely to experiences in Africa, smallholder farming.

> ***Tropical organic products***
> Arabica and Robusta coffee, instant coffee, tea, cocoa, vanilla, cotton, sesame, chillies, groundnuts, cashew nuts, honey, dried herbs, essential oils, shea butter, spices, cane sugar, fresh bananas, pineapple, mango, papaya, citrus, canned and dried fruits, fruit salads, fruit juice and concentrate, off-season vegetables, coconuts.

The author's background is in (sub)tropical agriculture with a specialisation in crop protection, and the early motivation to work in Organic Agriculture was from that professional interest side. However, very soon it was discovered that in conversion projects, especially in not too intensive systems, pest and disease problems are a minor thing and that Organic Agriculture is more about supply chain management, certification and financing.

The author has had the privilege to work with many tropical crops - the exceptions may be oil palm and rubber - and to have been confronted with the problems that these crops have in the organic situation. The work has always been with the exporter, or rather an agronomist-type person appointed by the exporter who is the counterpart, the person to be trained and who would in his/her turn work with the farmers. The search for solutions always is with the project staff and the farmers. Research stations, sometimes universities are visited for information. Back then, and even today they contribute substantially to problem solving by providing information but not by research.

In the below it is explained how a range of technical problems in a number of tropical crops were solved. The pest and disease problems appear to be the easy part. In the author's opinion there are two bigger issues, or work areas, that are more difficult to tackle. One is the lack of work on long-term soil fertility, the other one the connection between farmers (problems) and the research community (solutions). They are big because they are critical to the opportunity/responsibility of Organic Farming to make a positive contribution to solving some of the problems of modern agriculture and to the livelihoods of millions of smallholder farmers in Africa. If we do not work on those topics we will fail in our ambition.

The author wants to apologise in advance for the impoliteness of any-many quick judgements, the harshness of some opinions. There are many people working with the best of intentions and with limited resources. They need more space.

This paper is a practitioner's view. There are few references; there is little documentation of this kind of work. The author is keen to address that in the future.

6.2 The usual intervention

When working for the private sector, things need to happen pretty fast. There are few exporters and importers that are going to wait for 4 years for a product to become organic; no one is going to pay for that, hence the tendency to start with organic by default situations and/or on clean land.

Intervention toolbox

Management assistance
Business development elements
Staff selection, staff and field officer training
Farmer mobilisation, extension techniques
Farm input schemes, seed money for farm inputs
Setting up of tree/crop seedling nurseries, woodlots
Technical assistance and agricultural improvements
Development and revision of Internal Control System
Preparation of certification to additional standards
Quality management and quality improvement
Organic certification procedures and issues
Cost share in initial certification costs
Market surveys and buyer contacts
Product development (incl. retail)
Development of second crops
Mediation in trade finance
Risk assessment of the chain
Product presentations/brochures
Participation in organic trade shows

At the start of a project, the exporter is made aware of what is needed to convert part of his/her business to organic. The need for a capable field staff is explained and the exporter assisted in selecting the right person. Most of the work in the field is done by persons with an agronomy background. There are not so good experiences with using former government extension agents; it works far better with recently graduated, motivated persons who are good both in working with farmers and with papers.

As can be seen from the box, a whole range of support interventions are provided as needed. Within the time and budget limitations, priorities are set. The amount of work done on technical issues, also called Technical Assistance, is quite limited. Far more attention is paid to getting certified, improving the quality and establishing market relations. Except for the cotton project in Benin, there have never been research budgets in the projects. Technical issues are to be solved 'along the way' by the advisor, the field staff, a few farmers. If there is a budget, it is not more but a few thousand euros or dollars for materials.

As part of the support, farms are visited, meetings with farmers held, potential problems identified and suggestions made how to solve these. In most cases it is not too difficult to come up with solutions, it is more difficult to implement these. This is done for example by lead farmers with support of the field officers. The field officers function as extension agents, researchers, internal inspectors and contact persons.

Generally, farmers are very afraid of pest problems; they identify it as their biggest single threat. However, they often cannot distinguish pest insects from beneficials. They also hardly understand the mechanisms behind an infestation. They insist that they must spray; otherwise the crop may be lost. It does not matter that in daily practice they normally do not apply pesticides (and when, often not correctly) on their food crops. Researchers and government extension persons substantiate these claims by giving long lists of possible pests and diseases. As a Westerner who comes to talk about Organic Farming you are often seen as a very dangerous man, everybody says that it is impossible to grow crops and keep animals without chemicals. It takes some effort to find a person who knows something about, for example, biological control or green manures and is willing to defend such approaches in the group to get some acceptance. The result is that you often end up with a substitute treatment just to keep them happy.

The most difficult crop undoubtedly was and is cotton. In African smallholder agriculture, it is also the crop with the most pests. After a while it is only the bollworms that one has to bother about (because natural enemies take care of the rest), but then there are quite a number of them (see below). Another difficult crop is bananas, because of the Sigatoka disease (*Mycosphaerella fijiensis*). More or less satisfactory solutions were found (for example in Ghana) with Citrex and vegetable oils, although with better weather recording and crop monitoring the efficiency and effectiveness of application could further improve. The research for best treatments always has to go on; and the project holders do not always do that. When revisiting old projects, it is not uncommon to see exactly the same recommendations in use as were given ten years ago.

In the early days, when searching for green manuring and biological control, often recourse was taken to literature from the colonial era. Towards the outbreak of WW II, the knowledge of non-chemical methods in tropical agriculture was in its heyday. Since then, research went into the chemical solutions, both fertilisers and pesticides. The know-how from the old days is still a very good starting point.

6.3 Selected research questions

From the work in Africa, a number of examples are taken to give an overview of experiences.

6.3.1 Pemba nginigini in Northern Uganda

In 1995 the author was asked to identify a location for a possible organic cotton project in Uganda. A tour of the country was organised. Many places were visited. Uganda has seen plenty of violence and internal strife, but nevertheless the successive governments had done their best to introduce Green Revolution ideas. Government officials maintained that farmers were using fertilisers and chemicals. And indeed, when we stopped on the roadside and asked a somewhat frightened peasant who was at work in his field whether he used chemical inputs he hastily replied that he did. When asking a bit further, the even more frightened farmer had to admit that he had not since many years. Farmers were generally not very able to explain any of the pest problems so even when the author knew various cotton pests, it did not easily become clear which ones are present in Uganda. The eureka came when on some market place farmers were explaining about some insect eating another insect. Happy that at least somewhere farmers knew about natural enemies, the location for the cotton project was fixed.

It turned out to be predacious small black ants of the *Acantholepis* family that provided the natural control of the cotton insects. Farmers could identify the right species by tasting a couple of them. They knew what the preferred living conditions were, where they could find nests and had developed various ways of splitting populations of black ants, using for example wooden boxes filled with straw and meat offal, to carry parts of a population into cotton fields where the ants then built a new nest under a pile of organic material. Some sorghum and maize were sown early in the cotton field to be, and the pests that came onto those plants attracted the ants to go out into the field to search for prey. By the time the cotton plants were germinating and attracting aphids, for example, the ants were there to go onto the cotton plants. Throughout the growing season one could always see 2 to 3 of these little ants patrolling on each cotton plant. Because this element of stimulating the black ants (nginigini in the local language) as predators for the cotton pests, the organic cotton was soon called pemba nginigini. Pemba is Swahili for cotton. It even appeared on t-shirts.

Scientists from Serere research station (responsible for cotton) were invited to come and visit. One Egyptian IPM specialist came and after sweeping with his net confirmed that most pests were there but far under economic thresholds; that there was a healthy mix of pests and predators. Other scientists only wanted to come when we paid for their transport and lunches. However, the project managed to install an organic cotton field at the research station. Even when the ants are endemic in a certain belt across Uganda only, nests could be established there. The cotton growing was very successful. However, the exercise was discontinued after one season because there was no research to be done on this topic, no programme to attach it to so no budget and no interest. By that time the Egyptian had left. The party line was that farmers should use insecticides even when these were hardly available at the time (1996).

The Egyptian scientist had come with the suggestion to plant okra around the cotton fields to attract the cotton stainers *(Oxycarenus spp)* away from the opening cotton bolls. In Northern Uganda okra (malakwang) is used both as a leaf vegetable and for the fruit and was therefore also an addition to the farmers' diet. Theoretically, the cotton stainers that assembled on the malakwang could be treated in early morning with a neem spray, for example, but that was never done; the farmers did not find it worthwhile to do that. There was in general a low incidence of cotton stainers in the organic fields, whereas in a nearby research plot the cotton was red from the stainers, although they had sprayed the cotton 4 times in the season with pesticides there.

6.3.2 Le coton biologique au Bénin

Cotton is the main income earner for the country and thus the government, with the assistance of the French, has a heavy hand in the sector. There is no such thing as an *Acantholepis* in Uganda. There is a long list of pests, from early season aphids via *Syllepta*, *Spodoptera* and *Diparopsis watersi,* to the bollworms like the Sudan bollworm *(Diparopsis watersi*), the pink bollworm *(Pectinophora gossypiella)*, the false codling moth *(Cryptophlebia leucotreta)*, the American bollworm *(Helicoverpa armigera)* and the spiny bollworm (*Earias spp)* to the late season cotton stainer *(Dysdercus spp)*.

There is a bigger pest problem when the farmers spray already early in the season, for example against aphids. The cotton plant is perfectly capable of overcoming the damage of the aphids. Other sucking insects are not dangerous either as long as their natural enemies are not sprayed away. This is for conventional but likewise organic farms. Neem (*Azadirachta indica*) has always been the preferred botanical insecticide.

Some effort was done to improve the habitat for natural enemies and to attract them to the fields, like with EnviroFeast®, a feed attractant for predatory insects. As fields are small and there is plenty of vegetation around, the effect was not enough to justify the expense.

Experimentation was done with pheromone trapping for monitoring the adult populations of three of the bollworm species in order to better time the application of neem. This initially aroused some interest among the farmers who had never 'seen' the adults before. In the end it was more practical to prescribe calendar sprays, as a community activity, all at the same time, village by village. The farmers - their wives - make the neem concoction themselves from seeds collected locally. Buying a ready product would be too expensive and is likely less effective as the % active ingredient cannot be guaranteed in the ready product. Besides regular applications with neem, the manual collection of caterpillars is important. This is done as a family exercise, with the children, after school (if there is a school).

Once there was an offer from IITA, the International Institute of Tropical Agriculture to develop the application of the NPV or nuclear polyhedrosis virus to control the *Helicoverpa* bollworm. They said they had the virus already in the refrigerator, what was needed was the development of the modalities of the application. They were asked to present a budget. The proposal involved an international scientist, two local assistants, cars and laboratory facilities and the budget was far more then the entire project cost. The NPV stayed in the refrigerator.

What was somewhat more helpful was a co-financing facility with the Institut National de Recherche Agronomique du Bénin (INRAB) to investigate the effectiveness of various botanicals besides neem, like fish bean plant (*Tephrosia vogelii),* derris (*Derris elliptica),* papaya leaves (*Carica papaya*), chilli pepper fruits (*Capsicum frutescens*), garlic bulbs (*Allium sativum*), bark of the African Mahogany tree (*Khaya senegalensis*), the bulb of the `oignon du mort' (*Urginea indica* or *altissima*), wild peppermint leaves (*Mentha piperita*), Spinosad or spinozyne, cow urine and natural soap. This was done on experimental plots. The design was very professional with many repetitions to achieve statistical significance but also very complex to maintain. There were problems with the maintenance, observations and reporting (scientific validation) and the author does not recall seeing a final, authoritative report on the experimentation. In the end the farmers continued doing what was the recommendation from the beginning, calendar sprays with neem seed concoction.

6.3.3 CBD on Mount Kilimanjaro, Tanzania

For a long time it was deemed impossible to grow Arabica coffee organically on the mountain because of the Coffee Berry Disease *(Colletotrichum kawahae)*. The official recommendation was 6 to 9 applications at 3-4 week intervals. Benlate (benomyl 50%) was the preferred systemic fungicide until it was withdrawn because CBD started to develop resistant strains. Captafol was removed because of environmental hazards. The farmers continued to spray with Penerox (copper 50% wp). However, the continuous application of copper fungicides caused a build up of Cu-levels in the soil as high as 300 ppm. It was feared that this was leading to higher nematode populations because it killed micro organisms that are antagonistic to the nematodes.

The solution was to be in breeding CBD resistant coffee varieties, like the Ruiru 11. Such breeding programmes take a long time and are typically undersourced. To be attractive, these breeding programmes focus not only on resistance but also on early and high production. That means that the plants are quickly exhausted and must be replanted much more often than before. In Latin-America some estates replant coffee each 8 years, while the good old Bourbon lasted 40 years, if not longer. There is a continuous discrepancy between the statements of the breeding scientists that the flavour is ok, and that from professional coffee cuppers who say that one should be careful. In any case, the release of the Ruiru 11 was announced already 8 years ago but until today, the farmers are still struggling to get some nurseries going with this variety.

The idea of organic coffee growing was rejected by farmers, researchers and authorities alike; it was not possible because of CBD. There was a strong social control to make sure that all farmers sprayed. Nevertheless, we came upon a number of farmers who had given up because of the high cost. These farmers reported that after an initial increase in CBD, the disease virtually disappeared from their coffee fields in three years time. That encouraged us to start an organic project. Although there was the intent to promote, perhaps apply antagonist microorganisms, this never materialised. There was no need. There has not been an important incidence with CBD or of leaf rust in the last 5 years. Agreed, this might be because the climate is changing; rainfall is much lower. Conventional farmers continue spraying once or twice a year. It would have been nice to upload the ecosystem with antagonists but it was a too long shot to do when nobody else is interested.

Talking about a changing climate, there are also changes in the general context of coffee growing on the mountain. The erstwhile profitable coffee farms are split up among the sons when the father dies. This has resulted in many farms that are too small to make a living from. Also, by the time the old man dies, the sons are already in the city having modern jobs and families. It is difficult to sell off farmland because the ancestors are buried there. This results in absentee owners employing caretakers and the farms actually become weekend farms. The coffee is grown under shade in a mix of coffee interplanted with plantain bananas, a very good combination. Traditionally, the men do the coffee and the women do the bananas. Since the coffee prices crashed ten years ago, the women have the upper hand. Plantain is an important food crop and it provides weekly cash when a bunch is sold off. Coffee production brings money only over a short period of the year, which the women always say the men are wasting. Unless coffee production is made very attractive, when farmers are pulled together; coffee is a thing of the past.

6.3.4. Antestia coffee bug in Rwanda

In Rwanda, the government still insisted in 2005 on annual spraying of the entire coffee crop with Dursban to control the Antestia coffee bug *(Antestiopsis spp)*. They have been in this habit for a long time and it frustrates the start of any organic coffee project. This old approach is maintained by the coffee authority even when it is known that infestation is much lower in maintained (open) coffee bushes, that there are egg parasitoids like braconid wasps that can achieve 60-90% control, that Antestia was controlled in the past with natural pyrethrum - which is produced in the country, that if the farmers had to pay for the treatment it would cost them 20% of their coffee income, that Dursban kills the bees and therefore honey production. Even when a cooperative of farmers united in requesting the authorities not to spray, it was not possible to make an exception for the organic project.

It appeared that this spraying exercise was about the only thing that the coffee authorities were doing, and they are hanging on to it for dear life. It is somewhat disappointing that a renowned European research institute like CIRAD is supporting such countrywide applications by sending every year the obligatory scientist to confirm that it is necessary to do the exercise, upon which the EU releases the funds.
The author did not see any Antestia in the field at the supposed time of spraying (which was 5 months delayed). For this and some other reasons, the organic coffee project was stopped.

6.3.5 Work on cocoa, West Africa

Ivory Coast and Ghana are the world's most important cocoa producers. Although a good scientist can likely mention 10 or even more problems in cocoa, there are by and large two problems, black pod disease and capsids, nowadays called myrids.

There are two kinds of black pod, *Phytophthora palmivora* and *P. megakarya*. The last is the bad one. The occurrence of black pod depends on the amount of rainfall during fruit bearing, because humidity promotes the spread of the inoculum. That is produced on the infected pods. Removal of the infested pods should therefore be the approach but in practice one sees that great efforts are made to spray while no sanitation is done whatsoever. Various fungicides are used, including copper, which is allowed -under restrictions- in Organic Farming. Farmers prefer the use of a systemic fungicide like Ridomil but unfortunately this leaves residues in the cocoa bean. It is still approved for use in cocoa, it is freely available; it is also used on fruits and vegetables. Spraying is during fruit bearing (= harvesting). Cocoa traders are getting very nervous about this as, for example, Japan has a zero tolerance policy on pesticide residues in cocoa. They start traceability projects; to identify and exclude the farmers that use Ridomil.

The organic recipe would be tolerant varieties, aeration of the plantation, regular sanitation, application of antagonist microorganisms and avoidance of the *Megacaria* areas till we know better. Because of the antagonist mechanism, no copper should be used in organic cocoa.

Keeping the plantation fairly open is an issue of form and maintenance pruning which, even when these countries are highly developed cocoa producers, is not something promoted in the extension messages that go to the farmers, and it is not done by the farmers. Nor is a technique known in coffee and tea to have two layers of vegetation: higher shade trees, a space and then the cocoa trees, which would allow better ventilation.

The use of *Trichoderma* to reduce the incidence of black pod is well known and researched. It is not implemented in practice. In Ivory Coast, while visiting a research station, a report was seen on research to isolate indigenous species of *Trichoderma* in cocoa and to test the various species in the field in their effect on black pod. *T. asperelle* and *T. virens* had been found to have up to 80 and 70% control. No research had been done on the longevity of the application of inoculum or under which circumstances the dominance of *Trichoderma* could be maintained. The technique to produce *Trichoderma* inoculum is known. It requires a simple lab. What was needed to produce enough inoculum for an organic project was a simple machine, costing ten thousand dollars, to separate the inoculum from the medium. It appeared to be a technology that was ready to be put in practice on a larger scale; in an organic project. However, we were told that implementing such a technique would require a major research project, involving the research station we were visiting, the researcher we talked to and lots of money. Any project has to be approved by people in high places first. That means a long running-up time and important costs. One is in the habit of getting kickbacks from the pesticide industry. There is no 'overhead' or 'representation costs' on *Trichoderma* technology that makes it attractive to decision makers. And no, we were not allowed to have a copy of the report. Until the money was there the researchers would keep the know-how for themselves.

The second main problem is the capsids, nowadays called myrids. There are two species, *Distantiella theobroma* and *Sahlbergella singularis*. It is a man-made problem. Until thirty years ago there was no problem (or, it was under the economic threshold). It was not until shade trees were cut down to increase production, and the use of fertilisers that produce succulent leaf growth, that these insects became a problem. The insects feed on the young leaf, build up a population, and by the time the trees are bearing the pods, attack the pods.

The organic recipe is managed shade, stimulation of natural enemies -particularly *Oecophilla longinoda* ants-, visual or pheromone population monitoring, and well timed spot spraying with neem to reduce initial populations. The capsids prefer the fresh foliage in areas with full sunshine, when there are larger open spaces in the shade canopy. Too much or too long direct sunlight on the cocoa canopy should be avoided. It does not need to be a heavy shade; that would indeed reduce production. The technique of spreading the *Oecophilla* ant throughout the cocoa fields has been quite well studied. However, when not handled properly, the ants bite the farmers when they are moved about. Farmers do not like that, farmers do not like the work; farmers prefer to spray something even when that costs money. But it is at least modern. In Ghana, the situation has been spoiled by the government's mass spraying programme, where in the organic situation the government pays spraying gangs to treat the trees with neem.

It is a task for the national research institute to monitor the pest population, if only to alley fears that the organic project is a national threat (no kidding). An international research institute is doing research on pheromone trapping somewhere else in the country. They could/should provide some traps to monitor the population in the organic project. This has yet to happen.

Spraying with neem (or anything) should be based on the presence of myrids in the plantation, and should target new foliage, but it is not. Many times farmers were asked to show the myrids; and they could not find them. Very few farmers know the insect, while it is a fairly big one. There may be an ideal time of application but as it is difficult to get the neem seeds, the get the mist blowers repaired or a new one, to get the spraying team together and the operational costs available, the application is sometimes weeks if not months later. The preparation, concentration and application of neem are quite well researched. Whether it is always done properly is another question. The cocoa research institute recommends 4 applications with neem, like 4 applications of a chemical in conventional. The average number of actual sprays in conventional is something like 1.3. In organic it is max 1 time, and there is no difference. If farmers had to pay it all, spraying 4 times would cost them 30-40% of the gross income from the crop. No smallholder farmer is going to do that. In short, in organic the handling, timing and the effectiveness of sprays is an issue as well. One sees a situation where all the techniques are there, but the organic recommendations are not implemented. Where have we heard of that before?

One solution is that the management of pests and diseases is done in group work or possibly by a village committee. This requires a certain level of farmer organisation, management capability and, last but not least, a financial construction. Those are all aspects where things tend to go wrong in many parts of Africa, if not from the beginning, then after a few years.

6.3.6 Pineapple, Ghana

There is a potential for West African countries like Ghana, Cameroon and Ivory Coast to supply the European market with organic fruits like pineapple, mango and papaya. There are reefer lines twice a week, which allow the produce to be sea-freighted relatively fast to Europe. In Ghana, most of the pineapple is grown by smallholder farmers that supply exporters, who sometimes have their own, larger farm.

Until 2002 pineapple farmers throughout West Africa were using calcium carbide to induce flowering. Only some large scale, mechanised farms used ethylene. Flower induction is important to get the right size of fruit, which is very important for a market that wants a fruit size of between 1.4 to 1.6 kilos. Without flower induction this often means that only half of the fruits can be exported. It is also helpful to plan production and exports, as all fruits reach maturity X days after flowering. Harvesting on the right day is very important if one wants to get away from air freight. In the case of airfreight, the fruit can be harvested quite ripe as it is in the shop two days later. In the case of sea freight, it should be harvested earlier because considerable time is needed to cool down the pineapple on board the ship. Last but not least, it is far easier to harvest the field in 1 or 2 times, rather than look for the ripe ones during a period of three months. In short, flower induction would be good.

Early on, the organic certification bodies deemed calcium carbide a material that should be allowed in Organic Farming. It was used. However, in 2002 the European Commission prohibited the use of calcium carbide. One says because Portugal was growing organic pineapple on the Cape Verde islands in greenhouses where flowering is induced by smoking, others say because an American fruit company submitted a dossier that argued that calcium carbide acted as a potassium fertilizer, that ethylene was pure and much better. Ethylene was already approved in the US for organic pineapple farming.

It took until 2005 for the European Commission to approve ethylene. But ethylene is applied in a large quantity of water (drenching) using boom sprayers, while the majority of the West-African farmers are used to apply calcium carbide with a backpack sprayer. The regulation seemed to favour Central American producers to the West Africans.

A research project was initiated by Coleacp to investigate how ethylene can be applied by smallholders. Once this was commissioned, none of the contact persons ever reacted to emails. Some time later it was announced that the answer was in small pallets impregnated with ethylene and that the technique was being patented by the Belgian university which invented it. It is as yet (2008) not available. One can wonder about the cost when the material has to come from Europe and how accessible it will be for smallholders.

In the meantime there continues to be an important shortage of quality (correct ripeness), affordable (sea freighted) organic pineapple in the European market. Ironically, this is partly filled by a few projects that are certified by one particular certification body that has agreed with one or two European import authorities that calcium carbide is equivalent to ethylene. So those farmers are using calcium carbide again. Projects certified by another certification body and/or exporting to other European countries cannot use calcium carbide.

In a project introducing Eurep GAP type certification in the Ghana pineapple industry, research was done into the cost and benefits of better management practices on pineapple farms (NRI-NRET 2002). It showed that practices differ considerably from farm to farm, that if one calculates the cost price there is great variation. Most of all, the project showed that cost price calculations are hardly done. Such a technique, to know the cost of production, is not known, is not taught, is not available for pineapple farmers and exporters. This was confirmed in another study some years later, comparing coffee and fruit exporters (FAO 2007). That also showed that the type of organization (cooperative, local or foreign entrepreneur, international commodity house) is not a determining factor for how much smallholder farmers benefit from the conversion. It is the skills of the management that does it. Without having access to self-assessment tools, like cost price calculation, it is very difficult to develop the business quality of any operation. The success of organic projects depends substantially on how efficient the business is organized. More needs to be done on this type of business research-support-development.

6.4. The great question; long-term soil fertility

While this is fairly well done on commercial farms, especially in the case of fruits and vegetables, this is a problematic aspect that is usually ignored in case of smallholder agriculture (which constitutes 80-90% of agricultural production in Africa). While generally the conversion to Organic Agriculture in a smallholder situation results in a more varied rotation, the use of some leguminous crops and limitation of erosion, there is effectively little done about crop residue recycling and hardly anything in terms of returning nutrients. African smallholders do not have the understanding, the labour capacity and certainly not the finances to do much more. There is no pressure from certification bodies to enforce practices that would better maintain long-term soil fertility. They are satisfied with the absence of synthetic fertilizers. Shifting cultivation is the answer to diminishing soil fertility.

There is a lack of education, of understanding of what farming is among the farming population. Nominally, farmers may have had primary education but the majority has difficulty writing their name, doing a simple calculation or making a drawing of a plant -with roots- or an animal. Most extension agents can't do more than quote how many grams of NPK a plant should receive per season, or what a farmer should spray. There is little understanding of biological processes in nature. While it is possible to see results in the case of pests and diseases control, changes in soil fertility take much longer to observe. It is therefore far more difficult to motivate farmers to do a better job in soil fertility management.

Uncertain land tenure is of course another reason why farmers are not investing much in their land.

6.4.1. Sugarloaf pineapple, Ghana

In a traditional pineapple situation in Ghana (shifting cultivation) a series of soil analyses were made of untouched primary forest, 10 year fallow/secondary forest, first and second cycle pineapple fields, and in a 7 year old pineapple field. Sampling and analyses were done by a local soil lab. The results are the same for all situations, except that the primary forest has a slightly higher organic matter content. In all other situations OM, N, P_2O_5, K_2O, S, MgO contents were low to very low. Ph was high but not too high, which is quite common in West Africa. Some trials were done with mulching and compost applications, which after four and two years respectively did not show any visible effects at all. The rate of compost application was probably too low but higher rates would not have been economical at all for the farmers. This is about farmers having an annual gross income from pineapples of U$ 200-300 a year. The experiments should have been done on a longer term but the exporter had no patience for such hobbying. He wants to export certified organic pineapples.

6.4.2. Cotton, Benin

We have only one project in Africa that is well researched for its long-term soil fertility; the organic cotton project in Benin (Witte 1999). That situation is representative for a lot of rain fed field crop production in sub-Saharan Africa. Work was done on the organic matter and nitrogen management with soil scientists of the Wageningen University using the NDICEA model (van der Burgt 2006). Obviously, modelling has its limitations as a lot of assumptions are made. Nevertheless it is a good tool to understand the impact of certain measures.

We studied the situation where there is a rotation of cotton, peanuts and maize with a two-year fallow. It is a fairly sandy soil, with a humus content of 1.8% in the top 20 cm and a C/N quotient of 20. Five different management patterns were described, data collected and fed into a model. The reference situation was the conventional practice with fertiliser applications of 50 kg urea/ha on the cotton and maize. In that case there are already serious problems with phosphate, potassium and sulphur. Options were removal of crop residues or reincorporation, and normal grass fallow or improved (leguminous) fallow. This was calculated for two rotations; 10 years. The result of the organic rotation with incorporation of the crop residues is shown in the graph.

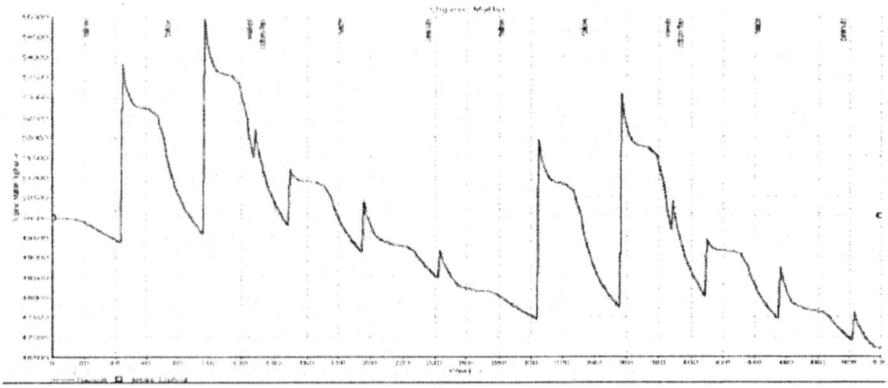

Figure 1: Soil organic matter dynamics with crop residue incorporation

It is only in the case when crop residues are reincorporated and an improved fallow with pigeon pea is established that the decrease in organic matter and nitrogen is reversed. However, it is customary to feed crop residues to animals, or to burn stalks on the edge of the field. Farmers (and their wives) find it also troublesome to incorporate crop residues when preparing the land. With cotton and maize there are also problems with the crop residues harbouring pests. The big thing is that we have never been able to convince farmers to invest labour and money in improved fallow, even when the leguminous crop provided pods for human or cattle consumption (like pigeon pea). Gross income from the cotton cash crop is € 200-300 per year for the average farmer. It is quite impossible that farmers voluntarily pay for rock phosphate or gypsum of which they do not see the effect. In short: there is a problem. There is a possibility that in many more organic field crop situations there is a steady decrease in soil fertility.

That the situation is often worse on the conventional side (even faster reduction of organic matter content) provides some solace but is not good enough.

6.4.3. Groundnuts and carbon, Tanzania

As it is difficult (= expensive) to educate-convince smallholder farmers to do a better job the motivation could be in ... money. The usual organic premium of 20% is not sufficient. However, in this age of global warming, soil organic matter is becoming an important tool to mitigate the effects of climate change. Apart from reducing erosion and by retaining water. reducing the effects of drought, it also plays a role as a carbon store. If one can increase organic matter, one could potentially sell carbon credits. In an organic export project (groundnuts) in Tanzania, the feasibility was studied of paying the farmers a premium for taking extra carbon fixing measures-better organic matter management, which investment would be repaid by selling the resulting carbon credits in 5, 10, etc. years time.

A baseline study has been made of the current farm practices and carbon stocks, and the potential to sequester soil carbon with improved farm practices. Over 90% of the carbon stock is present in the soil; the remaining is present in the sparse vegetation. The current farm practices of annual ploughing and burning of crop residues and the fallow vegetation for land preparation result in a steady decline of soil carbon and lead to land degradation. Improved practices, including reduced tillage, use of compost, mulch and improved fallow, can reverse this trend and stabilize or even improve the soil organic carbon content.

This is a project with 700 farmers. They could fix in between 9,000 to 21,000 tons of CO2 (depending on the practices used) within a five-year period. With a carbon credit price of 10 €/ton of CO2 this would be worth 90,000 to 210,000 €. This translates into 250 € per farmer. That is an interesting extra income. Besides that, the project (= the peanut exporter) would provide seeds and tools. As this is a pilot project there are other costs involved. As carbon fixation in agriculture is not so well studied, a lot of scientific evidence needs to be collected. There are of course also financing costs, as a bank has to provide the money upfront. As there is already an existing (organic) certification and premium system, the additional costs of carbon credit certification and of paying out an organic matter management premium are low. The project cost is € 275,000 while the outcome is not certain. No donor has been found to finance the project (Bodnar 2007).

Interestingly, planting alley trees has a faster carbon sequestration effect than growing green manures.

6.5. Motivating farmers

Although there are agronomic research questions, it is not difficult to come up with good advice for what the farmers should do to improve their farming, and livelihoods. The problem is that there is often quite a discrepancy between the advice and what farmers do in practice. This was a problem with the Green Revolution (if the technology was properly applied there would be a lot less problems with it); it is the same in Organic Farming. One can become desperate when one sees that good advice is not followed up; there are obviously many reasons why the farmers are not responding as the advisor wants them to. There are many barriers to adoption of new techniques, to change. In some societies there is peer pressure not to try to be better than the rest. In other situations, when you do better, the whole family or village descends on you to share the extra income. There is a labour side to it and with that a communication issue in the household. There is often an initial distrust of all things new, there are many histories of great new crops and ideas being introduced which were not properly followed up. Farmers are often cheated by middlemen or by their own cooperatives. The 'How to Get Farmers to Adopt Better Practices' issue is a big challenge.

In any community there are always a number of leaders who pick up on new things. In most organic export projects, you want all the farmers to go organic, from the leaders via the followers to the laggards, from the bigger farmers to the single widow. Often entire communities are converted; anyone not participating is a risk (to organic integrity).

In most African countries, the government extension service is pretty dysfunctional, in some it has disappeared completely. That leaves extension to those who want to sell something (the agrochemical salesmen), or to the ones that want to have something (the buyers), as is the case in organics. We have yet to work with a Ministry of Agriculture on the ground with Organic Farming, for the time being we talk about policy making.

In an organic export project, the foreign advisor trains and advises the agronomist of the project, of the exporting company, who is doing the extension work (apart from some other responsibilities). This field officer is a critical function; it is an understatement to say that not everybody is good at it. When this person is not motivated; one cannot expect the farmers to be. In some countries local advisory capacity is built, which allows for a back up, a further development of already established organic export projects.

A main motivating factor for the farmers is of course, the organic premium. This is first of all linked to not doing certain things (use of synthetic pesticides and fertilisers). The part of what they should do (redesign their farming system) is usually in the form of recommendations. It is already quite something to enforce good weed management as a condition for being certified. The projects that focus on achieving change do much better than the projects that are about getting certified. The bulk of initiatives in Organic Farming in Africa are the latter.

Research was done in some projects into what motivates or would motivate the farmers to be organic, or become better farmers. In some cases, Rapid Rural Appraisal exercises or baseline studies are done to try to understand better what the farmers' problems are, what they want, what they would be willing to do. Lists of incentives were drawn up. In few projects only it is analysed after say 5 years what worked and what did not work (Gibbon & Bolwig 2007). Research into the impacts of Organic Farming usually brings interesting and sometimes surprising results. It should be repeated regularly. How did the farmers' community experience the organic project after five years, ten, twenty years?

In the past 20 years all kinds of extension techniques have been tried. From the more top-down approaches, train and visit (instructions), to group meetings (let us all do this), supply of pruning equipment, to softer techniques like organising farmer to farmer excursions, to installing demonstration farms and organising meetings there, to group work in each others fields. We have also worked with traditional leaders, opinion leaders, with training of farmer-extensionists.

We have read about and sometimes seen in practice new developments in the field of agricultural innovation and knowledge sharing like Participatory Technology Development, Action Research, Farmer Field Schools, Local Agricultural Research Committees (Ayenor 2007). These were all on the research level; it is very seldom that these are employed as an across the board extension approach. These techniques all aim to improve adaptation of new practices by the farmers but generally they are very expensive to implement and difficult to maintain. One can visit areas where a few years ago there was a project with FFS, or a LARC, and the farmers hardly talk about it anymore and have gone back to their old ways.

Adoption of something new is a social process. In that respect it is amazing, also in our own daily practice, how easily the gender issue is forgotten. We continue to talk to the men while the women do at least part, if not most of the work. Man and wife do not always talk that much to each other (about farming practices), the wife does not always share in the income of the cash crop, etc.

Much more needs to be done about the question of how to convert communities to organics, how to achieve active Organic Farming, vibrant rural economies. There is only so much one can do when the advisor's budget is limited to a few visits of a few days during the first years. Any organic conversion needs a sort of continuous support.

6.6. The way forward

In Africa, there are all kinds of organic initiatives, small and large, export or local market oriented, started by farmers, exporters, professors or local NGOs, sometimes foreign development agencies. It can be in old and new cooperatives, in self-help or church groups, smallholder communities, commercial farmers, plantations. There is a lot already that can be analysed, lessons learned from, and research questions based on. There is plenty to do.

There is not too much communication between the different initiatives because roads are bad, physical meetings expensive, people perhaps culturally not used to share information. Research should be linked with the national organic movement, and research and academia are to play a part in getting that off the ground and maintained.

There is also the necessity of having a pan-African or at least regional organic research network. There is a lot to share, to learn from each other, to direct the efforts.

It has been said before but for research to be useful, it must be done with the farmers. Most problems need hands-on, practical fieldwork; simple and cheap solutions that are appropriate for the farming situation. From day one, the sustainability of any solution should be kept in mind. There are a few issues like long-term soil fertility and breeding that need long-term, fundamental approaches.

There are clearly identifiable technical problems, like a pest here or a nutrient shortage there. The socio-economic aspects are more complex, interdisciplinary, and deserve as much, if not more attention Any involvement of a research entity with a farming situation should be with a long-term perspective. Seasons are different over the years; the situation develops, so the technology applied has to change as well.

As there are plenty of sceptics there should be baseline and impact studies. Many of the people in town do not really know the farming situation. Things like open days, seminars on the research efforts are important. Too much of the knowledge is not accessible; some scientists do no want to share without being involved in any follow-up research. As research needs to be done with the farmers, most of the research effort must be done by local researchers. It would not be a bad idea, though, to link up national research with international bodies in order to get the international exchange.

Certification bodies have the responsibility to question the sustainability of organic projects. However, very few projects have the funds to pay for research, nor do development agencies want to invest in it. It might have to be a separately funded, flanking activity not harming the economic feasibility of Organic Farming but improving it.

There is a lifetime of work still to be done.

References

Ayenor, G. (2007) Assessing the effectiveness of a Local Agricultural Research Committee in diffusing sustainable cocoa production practices; the case of capsid control in Ghana. Int. Journal of Agricultural Sustainability, **5** (2&3) 109-123

Bodnar, F. (2007) Baseline study of carbon stocks and potential for soil carbon sequestration in organic crop production in South West Tanzania. Agro Eco, Netherlands

FAO (2007) Increasing incomes of smallholder farmers from organic export agriculture in sub-Saharan Africa. Rome (draft manuscript)

Gibbon, P. & Bolwig, S. (2007) The economics of certified organic farming in tropical Africa: a preliminary assessment. DIIS Working Paper, Copenhagen

Kotschi, J. & Müller-Sämann, K. (2004) The role of organic agriculture in mitigating climate change – a scoping study. IFOAM, Bonn

NRI-NRET (2002) The cost of compliance with a code of Good Agricultural Practice, the Ghana pineapple case. (Internal project document) Greenwich, UK

van der Burgt, G.J.H.M.; Oomen, G.J.M.; Habets, A.S.J. & Rossing, W.A.H. (2006) The NDICEA model, a tool to improve nitrogen use efficiency in cropping systems. Nutrient Cycling in Agroecosystems, **74** (3), 275-294

Witte, R. (1999), Towards sustainable soil fertility in organic, cotton based farm systems in Benin. Internal discussion paper, Obepab-Agro Eco, Netherlands

Review:

Development and Current Situation of Thai's Organic Vegetables

*Suchila Techawongstien *[1] and Somchai Chuachin[2]*

**Correspondence, E-Mail: suctec@kku.ac.th*
[1] Department of Plant Science and Agricultural Resources, Faculty of Agriculture, Khon Kaen University, KhonKaen 40002, Thailand.
[2] Bureau of Agricultural research and development 4, Department of Agriculture (DOA), Ministry of Agriculture and Cooperatives, Khon Kaen 40002, Thailand.

7.0 Summary

Thai organic production was initiated by farmers and local NGO's during the 1980's after evidence of health and environmental problems occurred due to the improper agrochemical use. Department of Agriculture (DOA) Ministry of Agriculture and Cooperatives, Royal Thai Government have officially started to promote organic agricultural products from 1992 on, by cooperating with rice trade companies. The Royal Thai Government first included sustainable agriculture in national policy in the 8th National Economic and Social Development Plan (1997-2001) as part of the "Sufficiency Economy" framework, and so-called "New Theory" farming. In general, Thailand's organic sector is driven mainly by government projects, private sectors, grower cooperatives, independent farmers, and especially nongovernment organizations (NGOs).There are many sectors of the Royal Thai Government Institute which have supported organic production in Thailand, i.e. Ministry of Agriculture and Cooperatives (MOAC), Ministry of Public Health (MOPH) and National Bureau of Agricultural Commodity and Food Standards (ACFS). In December 2004, the "National Assembly of Organic Agriculture" was held in Bangkok to promote Organic Farming; to improve standardization and to develop accredited certification agencies at the local and national level. Green Net and Earth Net Foundation estimated that the certified organic area increased from 2001 to 2005 (>900%). The number of farms increased commensurately, with 2,498 organic farms, representing 0.049% of the total number of farms in the country in 2004 (5.1 million farms). The Departments of Agriculture and Agricultural Extension, in response to Government policy on agricultural chemical-use reduction, have designed and extended "Good Agriculture Practice" (GAP) for major economic crops including vegetables. Department of Agriculture (DOA), Ministry of Agriculture and Cooperatives have classified organic vegetables into four crop types. They are 'organic' crops, 'non-toxic' crops, 'hygienic' crops and 'safe vegetables'. However, consumers normally recognize them in three distinct types, i.e. organic, no-pesticides (pesticide-free), and the 'safe-use-of-chemicals' vegetables. Thai's Organic Farming received the certification "Organic Thailand" from the DOA, based on IFOAM accredited certification agencies. In 2006, USDA Foreign Aricultural Service estimated that the total market value for Thai organic products was approximately US$20 million based on continued economic growth at a steady rate, which is about 0.12% of total agricultural exports. For overviews, Thailand's organic exports have a bright future due to excess of demand over supply in international markets. At present, the biggest customer of Thai organic produce is the European Union, especially Germany, the United Kingdom, and France; while Japan and Singapore are annually increasing.

7.1 The development of organic vegetable in Thailand

Organic Agriculture in Thailand was established by farmers and local NGO's during the 1980's after the health and environmental effects of improper use of and heavy reliance on agro-chemicals began to manifest themselves (Phinthupan, 2000; Kramol et al., 2006; USDA, 2006). However, the emerging popularity of Organic Agriculture in Thailand has resulted from a combination of three major trends (Ellis et al., 2006). The first trend is an increasing public awareness of healthy living. The second trend is the development of sustainable agriculture in response to the crisis faced in the farm sector. The last trend is the rise of environmental awareness, starting from a concern for environmental protection and conservation, but later transforming into a broader agenda covering the impact of conventional agriculture on environment, ecology and biodiversity, including land use, landscape, biodiversity, and pollution caused by use and mis-use of agro-chemicals. The convergence of these trends contributed to the pace of development of Organic Agriculture in Thailand during the early 1990s until the present day.

Department of Agriculture (DOA), Ministry of Agriculture and Cooperatives and Royal Thai Government have officially started to promote organic agricultural products from 1992 on, by cooperating with rice trade companies; Nakorn Luang Co. and Siam Wiwat Co. for exporting to Italy (DOA, 2000). Meanwhile, Thai government, Thayang Agricultural Cooperative started the cooperative project with ToTo Consumer Cooperative, Japan for exporting banana to Japan. During 1999-2003, the Department of Export Promotion, the Ministry of Commerce in coordination with the Ministry of Agriculture and Cooperatives, the Thailand Institute of Scientific and Technological Research and the Kasetsart University set up the pilot project of organic products for export. This project aimed to produce 6 kinds of vegetables; asparagus, babycorn, small banana, pineapple, okra and ginger, and export through many exporting companies to Singapore, Japan, Europe and USA. Recently, the Royal Thai Government first included sustainable agriculture in their national policy in the 8th National Economic and Social Development Plan (1997-2001) as part of the "Sufficiency Economy" framework (NESDB, 1997), which aimed to transform 4 million hectares (or 20% of the total of farming area) into more sustainable agricultural practices - covering natural farming, Organic Agriculture, agro-forestry systems, integrated farming systems and so-called "New Theory" farming (Kramol et al., 2006). A few initiatives supported by the Thai Government included a grant about US$ 20 million to the Sustainable Agriculture Foundation Thailand to coordinate and implement sustainable agriculture and natural resource management in 19 agro-ecosystems throughout the country (Panyakul, 2001; Ellis et al, 2006).

In addition, the recent pressures from health and environment-conscious consumers for safer food and increasing global competition, have forced the Government to reinstate the policy of sustainable agriculture towards protecting the environment and stimulating the rural economy. In December 2004, the "National Assembly of Organic Agriculture" was held in Bangkok to promote Organic Farming; to improve standardization and to develop accredited certification agencies at the local and national level. The first stage was to reduce the use of agricultural chemicals including insecticides, fungicides and herbicides, and to provide financial support for the construction of 7,000 organic fertilizer plants nationwide.

7.2 Production: what, where and how much

In 2005 the Green Net and Earth Net Foundation reported that the Thai organic market was estimated around 29,415 tons, valued at US$ 23 million per year. However, Thailand has considerable potential in producing processed organic products, both basic (fresh fruit and vegetable) and processed, which will increase the added value in both domestic and export markets. Dried fruits, canned fruits and vegetables, fruit and vegetable juices, and processed cereal foods are of greatest interest and market potential in this respect.

7.2.1 Diversity of production systems

In Organic Agriculture, two lines of production systems are operating with different approaches (Unno, 2003). The first is based on a sufficiency economy - emphasizing household food security, food safety and income stability. This farming system tries to mimic natural ecosystems with poly-culture of local vegetables, herbs and medicinal plants and fruit crops in a home-garden system. The result of this 'integrative production' design is to provide food and income in a more sustainable manner. NGO initiatives on alternative agriculture advocate this thinking, seeing Organic Agriculture as a means to improve rural livelihoods, and advocating such local food systems as community-based economic development. This type of production is often organized through groups and networks. Certification from the authorized agencies is regarded as essential, but may arguably be not so necessary with the production being mainly for local markets. The contrasting approach is a more market-driven organic production system -as advocated by private exporters, processors and food-exporting companies who see the potential of organic food products in more affluent markets particularly in Europe, North America and Japan. Contract farming has been encouraged as a result between private food exporting companies and farmers to produce selected crop commodities, especially for sweet corn, French bean, asparagus, lettuce and egg-plant. Production continues to be a monoculture system, where production inputs include organic fertilizers and herbal or plant extracts for pest management.

The Departments of Agriculture and Agricultural Extension, in response to Government policy on agricultural chemical-use reduction, have designed and extended "Good Agriculture Practice" (GAP) for major economic crops including vegetables. GAP is analogous to the 'safe-use-of-chemicals' practice. The practice is based on the concept that production is agronomically feasible with carefully selected chemicals for pest management; financially manageable for the farmers with promising economic returns, and that the products are affordable and safe for consumers.

Concerning health and environmental quality, the Institute of Organic Crop Production of the Department of Agriculture, Ministry of Agriculture and Cooperatives (MOAC) has classified four types of crop production systems. These are 'organic' crops, 'non-toxic' crops, 'hygienic' crops and 'safe vegetables'. However, consumers normally perceive three distinct types of "safe" vegetable products, which are organic, no-pesticides (pesticide-free), and the 'safe-use' of chemicals product category (Kramol et al., 2006).

Since the demand for organic products is steadily increasing, private farms have invested in Organic Farming for the selected production of fruit crops and vegetables. These specialized farms seek their own market outlets and receive the certification "Organic Thailand" from the Department of Agriculture, based on IFOAM accredited certification agencies (Table 1).

Table 1: Registered and certified Organic Agriculture production (hectares) in Thailand in March 2004 (Department of Agriculture, 2005)

	'Organic Thailand' Certificate			Organic Agriculture Standardization Registered*		
Region	vegetables	vegetables and others	others	vegetables	vegetable and others	Others
North	98.24	75.68	395.44	0	116.16	10.08
Centre	55.04	1402.24	52.88	5.68	110.55	261.14
Northeastern	3.76	102.24	1773.04	9.6 1	497.88	2689.40
South	0	0.80	88.80	0	5.60	0
Total	157.04	1580.96	2310.16	15.28	1730.19	2960.62

Source: Kramol et al., 2006.

7.2.2 Production groups

In general, Thailand's organic sector is driven mainly by government projects, private sectors, grower cooperatives, independent farmers, and especially non-governmental organizations (NGOs). The production systems are operated by different major groups or institutions with different purposes and operating systems. The five following are the common organic production groups or institutions in Thailand (Kramol et al, 2006; USDA, 2006; Ellis et al, 2006).

The Sustainable Agriculture Foundation was established by a network of non-government organizations (NGOs) promoting Organic Agriculture and community-based natural resource management for sustainable rural livelihoods throughout Thailand. Several local NGOs working with rural communities have developed Organic Farming practices for the production of vegetables, fruit and rice. The most important of these are Green Net Cooperative Ltd, Lemon Farm Cooperative and Mae-Tha Sustainable Agricultural Cooperative.

Private agri-food companies are producing on a more industrial scale "safe" vegetable for export markets. Private sector producers may be divided into three broad categories i.e. producers with single farms (traditional family farms), the large-scale corporate farms and the smallholders working together as a grower group or under a common organic project which is becoming increasingly common today. The common arrangement in the last category includes credit support for material inputs including seed, fertilizers and other chemicals. The usual vegetable crops grown under contract include soybean, French bean or snap bean, sweet corn, baby corn, cucumber and purple eggplant in monoculture systems. These crops are produced under the company's GAP, which tries to reduce production risk through the use of chemicals selected by the company. Farmers have no alternative growing systems to choose. The major companies are Chiang Mai Frozen Food Plc., Lanna Agro Industry Co. Ltd., Exotic Farm Produces (Thailand) Co., Ltd., Rai Plook Rak Organic Farm, Rangsit Farm Co., Ltd., River Kwai International Food Industry Co. Ltd. Thai Organic Agi Co., Ltd. and Thai Organic Product Co., Ltd.

The Royal Project Foundation is in the process of transforming chemically-based enterprises to more 'safe-use' and organically managed production as farmers (mainly in the Northern area) gain more farming skills and experience. The project size differs by location, but each farmer is given approximately 0.16 ha to farm organic vegetables and herbs. The crops are then sold in the local market at prices equal to "conventionally grown" vegetables.

Most families have livestock and rice as well, but from vegetables and herbs alone families are able to earn an average about US$ 73.5 per month. Although the impact on the market is small, the project does propagate the idea of consuming organic goods and has been successful. Its products are commonly distributed under the brand name "Doi Kham". Production of organic vegetables in a few key villages is now operating with selected vegetables. For instance, Ang Khang village has a registered production area of 38.56 ha and has received certification under "Organic Thailand".

The Santi Ashoka Group is a religious group of communities, who have chosen self-sufficiency and 'living in harmony with nature'. The organic products produced are marketed within the community network across the country. The group also creates its own food stalls to serve the general public at an affordable price. The product quality is high and so is the demand, simply because of faith and trust of the consumers. The group is considered highly successful and sustainable without intervention from either government or non-government organizations. It believes that the 'inside out' approach involving a paradigm shift in mind and soul is a necessity for such a change.

Independent farmers, who intend to convert to pesticide-free production systems, have collaborated with district extension offices to experiment on pesticide-free farming practices. The progress to date is not encouraging, as there are no formal marketing arrangements.

7.2.3 Land area and number of farms under organic production

In general, the organic land area in Thailand is categorized into certified and non-certified, there has been however, no systematic record of organic production in Thailand, and it is therefore difficult to evaluate the situation precisely (Ellis et al., 2006). The Green Net and the Earth Net Foundation stated that the certified area under Organic Farming increased from over 2,100 ha in 2001 to 21,701 ha in 2005, representing 0.10% of the total agricultural land area (21 million ha). The land area devoted to Organic Agriculture increased by over 19,500 ha from 2001 to 2005 (>900%). The number of farms increased commensurately, with 2,498 organic farms, representing 0.049% of the total number of farms in the country in 2004 (5.1 million farms).

7.2.4 Volume and value of production

Although the largest organic production category is organic rice, fresh vegetables and herbs are the second ranged. Organic vegetables are mainly leafy vegetables, especially the salad type and Chinese vegetables, produced mainly in central Thailand and in the North, especially for the Chiangmai province. Somehow, many organic vegetable farms also produce organic fruits (Table 2).

Table 2: Production and market value of organic vegetable during 2003 – 2005

Crop	2003		2004		2005	
	Production (tons)	Value (m Baht)	Production (tons)	Value (m Baht) [1/]	Production (tons)	Value (m Baht)
Vegetables and herbs	2,671.28*	160.28*	2,656.73	159.40	4,618.18	255.83
All organic produces**	9,756.05	375.13	15,966.08	608.79	29,415.10	920.36

* including fruit; ** including rice, field crops, fruits and others organic crops

[1/] 35Baht ~ 1US$ or 50Baht ~ 1Euro in 2006

Source: Adapted from Ellis et al., 2006.

7.3 Government Policy, Regulation and Certification

7.3.1 Government Policy and Regulation

There are many sectors of which the Royal Thai Government Institute has supported organic production in Thailand, i.e. the Ministry of Agriculture and Cooperatives (MOAC), The Ministry of Public Health (MOPH) and the National Bureau of Agricultural Commodity and Food Standards (ACFS) (Table 3). Regarding the report of Ellis et al., (2006), Thailand's National Agenda on Organic Agriculture was launched in October 2005. The 5-year programme is aimed at supporting 4.25 million farmers (0.85 million in 2006) to use organic inputs instead of agro-chemicals on an area of 13.6 million ha (2.72 million ha for 2006), reducing the total import of agro-chemicals by 50% as well as boosting organic exports by 100% annually. The programme aims are to be achieved through various supports and intervention mechanisms, including seminars, training, general promotion, and setting up organic fertilizer factories. There are 26 agencies from 6 ministries involved in the programme, coordinated by the Land Development Department, Ministry of Agriculture and Cooperatives (MOAC). A US$ 31.5 m budget is allocated for this programme in 2006. The main aims are to reduce the use of agricultural chemicals and to promote Organic Farming in selected provinces as pilot projects. Key technological innovations include soil improvement practices with green manure crops and composting and biological pest control measures. A compost pellet-making machine designed by the Thailand Institute of Science and Technological Research has helped disseminate the use of dry compost, either by broadcasting or side-dressing application. Furthermore, the Cabinet has set up a national Organic Agriculture committee to advise the government on organic agricultural policy development. Practically, organic policy-making is a top-down process with relatively few mechanisms for consultation with the wider stakeholder community.

The organization responsible for implementing the regulatory framework for Organic Agriculture is the National Bureau of Agricultural Commodity and Food Standards (ACFS). ACFS consists of 5 divisions - Division of Agricultural Commodity and Food Standards Policy, Office of the Secretary, Office of Commodity and System Standards, Information and Technology Center, and Office of Commodity and System Standards Accreditation. ACFS drafted National Standards for Organic Agriculture, including production, processing, labelling, and product sale. The draft was finalized and approved by the Cabinet in 2003. The ACFS standard is similar to CODEX and IFOAM standards, in that it sets standards for standard-setting organizations, rather than standards for production, i.e. they are not intended for use by certification bodies conducting inspection and certification. At present the standards are still voluntary. In August 2004, ACFS received the first application for organic accreditation from the Organic Agriculture Certification Thailand (ACT). Accreditation was granted in mid-2005.

The Ministry of Public Health (MOPH) declared 2004 as the "Year of Health for All", and staged a public campaign for consumption of so-called 'clean and healthy food' as a way of overcoming some health problems (Kramol et al., 2006). Interwoven issues of health, environment, agriculture and the food system have been brought together by key institutions within the MOPH, notably Health Systems Research Institute and the Health Systems Reform Office, in cooperation with the Thai Health Promotion Foundation (Table 3). Many public forums had been organized to make the public aware about the impacts of agricultural chemicals on the environment, human health, and social costs. In fact, the MOPH and affiliated institutions play an important role in increasing public awareness of the potential hazardous effects of agricultural chemicals, and directly try to create an increasing demand for pesticide-free and organic agricultural products. The MOPH have drawn up a strategic plan for national food systems, with the aim of shifting Thailand towards becoming a country of greater food safety. The MOAC remain responsible for setting up systems to oversee proper production practices through the use of GAP, so that better vertical integration from farm to table of more safe and quality food can be achieved.

Table 3: Key Royal Thai Government Institutes supporting organic production

Institution	Technology services
Land Development Department (LDD)	Establishes farmer volunteers or soil doctors to help soil diagnosis Distributes composting technology, microbial culture, green manure crops
Department of Agriculture (DOA)	Introduces biofertilizer technology, integrated pest management Designs "Good Agriculture Practice" for main economic crops
Department of Agricultural Extension (DOAE)	Provides training in biological pest control, promotes the use of Trichoderma spp. for controlling some soil-borne disease
Thailand Institute of Science and Technological Research (TISTR)	Develops machine for pressing dry organic compost into pellets

Source: Kramol et al. (2006)

7.3.2 Certification

The Organic Agriculture Certification Thailand (ACT), which is a private certification agency founded in 1995, provides professional organic certification services for all farm production as well as the processing and handling operations (Panyakul, 2003). For pesticide-free agricultural products, the farmers can approach the local agricultural extension office, or the Regional Office of Agricultural Research and Development (OARD) for tests of chemical residues. However, these tests can not detect heavy metals in vegetables. In addition, the Provincial Public Health Office also conducts a periodic analysis for chemical contamination in agricultural and food products, particularly of fresh vegetables that are sold as pesticide-free and organic. By Febuary 2002, the Institute of Organic Agriculture of DOA has developed its own standard known as "Organic Thailand" to provide certification for individual growers, farmer groups and private companies. This certification lasts for one year and is renewable on a yearly basis.

Ellis et al. (2006) reported that there are 3 groups of certification bodies offering organic certification services for producers in Thailand. They are Thai government bodies, Thai private entities and foreign entities.

7.3.2.1 Thai government bodies: Public certification initiatives

In order to secure international acceptance and recognition, Thailand developed national standards for production of organic products, to conform to existing international standards. The standards were drafted in 1999 by the Thailand Institute of Science and Technology Research (TISTR) in cooperation with the Department of Export Promotion, Ministry of Commerce. The final version was adopted on 18 October 2000. In 2002, the DOA established the Institute of Organic Crops as a national certification body, and also to support research and development into organic crops. By February 2002, DOA had established the "Organic Thailand" brand and established 5 pilot projects producing 15 organic crops, managed in collaboration with farmers, individual experts, the private sector, and consumer groups. In September 2004 there were 440 DOA-certified farms covering 1,763 ha.

7.3.2.2 Thai private entities: Organic Agriculture Certification Thailand (ACT)

This certification method is based on collaboration between growers and consumers to accept organic products; and organic farms from the Chiang Mai Province are an example of these entities. In general, the Organic Agriculture Certification Thailand (ACT) is an independent private certification body, which was established in 1995. Its members are producer organizations, consumer groups, NGOs, environmentalists, and academics. ACT was the first and is still the only Thai-owned organic certification body offering internationally-recognized organic certification services. ACT has established its own standards for crops, wild product harvest, processing and handling but its current certification covers only organic rice, vegetables and bean products. ACT operates throughout Thailand except for the southern region, and received ACFS certification in 2005. ACT has forged a number of international linkages. In 2002, ACT was the first certification body in Asia to receive IFOAM accreditation, with the support of the International Organic Accreditation Services (IOAS). Its organic standards are recognized as equivalent to IFOAM (International Federation of Organic Agriculture Movements) Basic Standards. Since then, the Swiss government has also recognized ACT, which later led to acceptance by global organizations such as BIOSWISS (Switzerland), BLIK (Belgium), KRAV (Sweden), Die Bio-Bauern (Austria), Ecocert (Canada), and Agriculture Biologique (France). As of June 2005, there were 1,413 farms certified by ACT, covering 4,752 ha.

7.3.2.3 Foreign entities: Foreign-owned private certification agencies

These entity certifications are different due to the import countries. Bioagricert (BAC), BCS Öko-Garantie GmbH and some other organizations like Soil Association (UK), IMO (Switzerland/Germany), OMIC (Japan) and Skal (Netherlands) are examples of foreign-owned private certification agencies for Thai organic produce.

7.3.3 Markets

USDA (2006) categorised the Thai organic market into four channels through which organic products are sold: supermarket chains, specialized shops, direct marketing (membership) and also some other producer groups, even though these sell at the locally small scale.

Organic products in "supermarkets" are often sold under the umbrella of "health" products. They are intermingled and placed on the same product shelves as non-organics. Only when there is a product promotion they are distinguished from their non-organic counterparts. Supermarkets generally focus on vegetables with fruits as a far second, though larger more upscale stores do carry a wide variety of organic products including cereals, tea, potato chips, and juices among many others.

The **"specialized shops"** focus on organic products, but include other products and foods considered "healthy". Another specialized outlet that carries organic products is OTOP stores. OTOP is the acronym for a government project called "One Tambon, One Product", which aims to promote a single product in each district. Since in some districts and provinces the Thai government is promoting Organic Agriculture, some organic products fall under the umbrella of this project and are marketed in OTOP stores across the country.
The **"direct marketing"** of organic products does occur, but it is limited and mainly includes fresh vegetables through co-operatives or membership groups.

For the **"other locally producer groups",** there are a small number of restaurants that serve organic foods. However, they are generally more geared towards "health food" than organics specifically. A restaurant may only have a few dishes made with organic products. Some resorts boast "organic" restaurants, some of which use produce from their own gardens, but this does not mean that all ingredients or dishes are made with certified organic foods. In addition, Thai organic markets comprise many components, i.e. current market, pricing structure, distribution channels, domestic perception and exports (Ellis et al., 2006).

7.3.3.1 Current market

In 2002 Thailand began to experience an economic recovery and the early signs of a revitalized domestic market. However, urban consumers were only becoming aware of the benefits of consuming organic food. This was partly due to the lack of available information to help consumers differentiate between organic and 'chemical-free' produce, which was also available in the market, and promoted by two separate government schemes. Meanwhile, most of the certified organic crop production at that time was export-oriented, leaving only an insignificant volume for the domestic market, i.e., fresh vegetables and grains – mostly rice and beans. Since that time, the range, volume and brands of fresh organic vegetables have all considerably expanded. By the end of 2004, many certified brands of organic farm produce appeared in local supermarkets and modern trade outlets, particularly in Bangkok. These new entrants led to an increasingly competitive environment and helped reduce consumer prices. Reliable sources of data on organic produce are hard to find. The situation is confused by the various standards or systems of certification for organically grown crops and produce. This made it difficult to categorically differentiate between the two market segments. USDA Foreign Agricultural Service (FAS) estimated that in 2006 the total market value for organic products will be approximately US$20 million based on continued economic growth at a steady rate. This is equal to 0.12% of total agricultural exports.

7.3.3.2 Pricing structure

In a study of 20 fruits and vegetables in Bangkok, Green Net has found that organic produce costs up to 133% more than conventional chemically grown produce and approximately 40%

more than "hygienic" produce of the same type. With this evidence, many farmers are struggling with the reality that they are not able to get a premium price for organic foods in local markets. Many customers will buy organic foods before conventional ones, but are not willing to pay more for organics because of their appearance or because they do not have confidence in organic claims. This has led some farmers to sell their organic products at conventional prices to schools and hospitals. According to FAS's observations, the price premium particularly on vegetable items is more than 50% in most supermarkets that carry organics in Bangkok. However, consumer surveys indicate that customers are willing to pay only 15-20% more for organic produce.

Despite the increase in volume and number of producers which have exerted downward pressure on prices, current markets for organic produce still favour producers as demand continues to outstrip the available supply. Organic produce on supermarket shelves formerly attracted a 20-50% price premium compared with conventional produce (Table 4). Surveys show that consumers are ready to pay a maximum premium of 10-15% for organic produce, on condition that the quality must be at least equivalent to conventionally grown produce. As mentioned, average prices for organically grown fresh vegetables sold in supermarkets in Bangkok continue to gradually fall, since the introduction of organic produce through modern trade outlets at the end of 2001. However, the price base is still higher than for 'hygienic' and 'chemically produced' vegetables. It is anticipated that in the next few years, when more producers have gone through the transition period of converting their farms, more organic agricultural produce will be launched in the market, leading to further downward pressure on prices.

Table 4: Price comparison of fresh vegetables in Bangkok supermarkets (US$/kg)

	Dec 2001	**Jul 2002**	**Mar 2003**	**Dec 2003**	**Mar 2004**	**Aug 2004**	**Feb 2005**
1. Conventional vegetables	1.18	1.21	1.15	1.69	0.87	1.03	0.88
2. Hygienic vegetables	1.61	1.77	1.04	1.67	1.36	2.45	1.42
3. Organic	2.60	1.97	1.94	1.89	2.26	3.98	1.54
4. Organic premium above hygienic vegetables (2)	61%	11%	46%	12%	40%	38%	8%
5. Organic premium above conventional vegetables (1)	120%	62%	69%	11%	62%	74%	43%

Source: Adapted from Ellis et al. (2006)

1) Distribution channels

The major products in demand are organic vegetables, and the supply is domestic. Rai Pluk Rak, Rangsit Farm, and Splendid brands are prevalent. Out of the vegetables that are available, many of them are leafy vegetables or "salad" and different types of lettuce and cabbage. However, many farms that produce organic products, which are usually uncertified, sell them in small local markets. There are very few organic products available other than vegetables.

Distribution channels for organic agricultural products in Thailand have considerably developed since cooperatives launched the organic market nearly 10 years ago. During the early years, consumers could only purchase health food products from small retail shops, although these numbers later proliferated. When it reached the peak of popularity there were more than 100 shops of this kind. However, the distribution structure changed radically once retailers recognized this trend. Soon organic and health foods became increasingly available in most modern trade outlets (supermarkets and hypermarkets). Some companies have even made the allocation of shelf space to organic produce a part of company policy. Currently, the main distribution channels in Thailand consist of 7 channels; depending on their different operating systems. There are member systems, weekly markets, occasional markets, retail health shops, health supermarkets, supermarkets and discount stores and organic restaurants.

2) Domestic perceptions

Thailand's organic consumers may be characterized as follows: urban and middle-class consumers, with 1-2 children and average income about US$ 880-1760 per family per month. Principal shoppers are housewives or female members of the family, who are more health conscious compared with their male counterparts. Consumers have access to data and information beneficial to their health and also to sources of organic food and health products, mainly from printed media, TV and radio.

For the organic market value, Green Net estimated that in 2004 it was about US$18 million. The domestic market for health food in general is estimated at close to US$75 million. There is a lack of comprehensive studies on the organic market. Even for retailers, organic sales is such a small percent of total sales that it is not tracked. USDA, Foreign Agricultural Service estimated that growth of the market will be approximately 10- 20% per year over the medium term based on economic indicators, valuing the market to about US$17 million in 2005 and US$20 million in 2006. As previously mentioned, increasing Interest and demand for organic farm produce among local consumers is due mainly to the fact that consumer concerns over chemical contamination and other hazardous substances present in farm produce and foods have often proved to be well-grounded. A flood of health foods and products entered the market to capitalize on these concerns. A product may carry multiple seals or labels from various government agencies or international organizations demarcating different types of certification, whether it is for organic, GAP standards, food safety or others. There is not a single label that signifies that a product has met organic standards because there is no overarching certification required. Organic products are subject to the same labeling requirements as conventional products (USDA report, 2006).

3) Exports

Thailand's organic exports have a bright future due to excess of demand over supply in international markets. Thai produce such as rice and tropical fruits and vegetables in particularly have a high demand. With its comparative advantage in production, Thailand is well-placed to serve the world market. Major importers of organic farm products include the European Union, especially Germany, the United Kingdom, and France. Demand in Japan and Singapore is also significantly increasing, with the Japanese market for organic farm goods expanding at the highest rate in the world. Among the most popular products are fresh and dried tropical fruits and vegetables, as well as processed agricultural products.

Thailand's comparative advantages lie principally in export of tropical products not produced in the developed markets, off-season products, in-season products where demand outstrips supply, and specialty or novelty products. It is important to note that the domestic and export markets can not be considered in isolation. The rapid development of the domestic market for organic food and products contributes to the overall stability of the sector by easing supply fluctuations and broadening the diversity of available produce, thereby supporting the export market. A mature domestic market also provides a ready market to absorb export surpluses and produce which falls below required export specifications. The Department of Export Promotion (DEP), Ministry of Commerce, undertakes a number of activities aimed at facilitating exports. Apart from supporting participation of exporters at global and regional organic events (e.g. Biofach, Germany and the Natural Products Organic events in Singapore and Hong Kong), DEP also organizes business incubation and training events to develop the export sector, and plans to develop an information website catering specifically for the organic sector.

7.4 Challenges and Opportunities

For Greennet and Earthnet points of views, Thailand's organic exports have a great potential due to fast growth in international markets, particularly in the EU member countries (Ellis et al., 2006). Thai organic produce (particularly rice and tropical fruits and vegetables) is in very high demand. With its comparative advantage in production, Thailand is well-placed to serve the world market. However, the stringency of current requirements presents barriers to participation of smallholders in both export and domestic organic markets. Training, education and support (especially for small farmers for certification and marketing), will all be important drivers of sectoral growth. Review and upgrading of Thailand's regulatory and control systems will also be required to ensure equivalence with EU standards, as a step towards Thailand's inclusion in the EU Third countries list. USDA (2006) finally evaluated the advantages and challenges for Thai's organic production in term for suppliers and also competitors, concluding that Thailand's organic exports have a bright future due to the excess of demand over supply in international markets (Table 5).

Table 5: Advantages and Challenges of Thai's organic produces

Advantages	Challenges
Organics becoming more popular in Thailand	Organic products not always available
Complementary U.S./Thai growing seasons	Higher transport costs and higher product prices
High end importers in hotel, restaurant, and spa industries- U.S. suppliers equipped to provide organic products that cater to patron's/ Western tastes	Thai consumers don't have an expendable income to pay for organics prices that are higher than non-organic products
U.S. organics products and standards recognized as high quality	regionally grown organic produce and herb varieties that are not grown in the U.S.
Thais are accustomed to locally and government heavily promoting "safety food"	Thai consumer confusion about the meaning of organic- multiple terms being used (hygienic, pesticide- free, safety food)
Market expanding- slowly	Domestic market small-niche
Health and environmental concerns increasingly effecting consumer food choices	With time Thai organics producers will be able to meet U.S. organics standards
Number of "Green Shops" increasing, larger supermarkets carry organics	Organic products not distinguished from non-organics on supermarket shelves
Thai consumers receptive to marketing and promotions	Lack of consumer confidence in organic quality because of a lack of marketing on the benefits of consuming organic products to end consumers
Some areas in Thailand have limited production capacity because of water and manure shortages	Thai consumers not likely to pay premium prices at U.S. levels-issue of profitability and competitiveness

Source: USDA (2006)

References

Agricultural Studies Department (2000) Standards for Organic Crop Production in Thailand. Ministry of Agriculture and Cooperatives

European Commission (2005) Proposal for a Council Regulation amending Regulation (EEC) No 2092/91 on organic production of agricultural products and indications referring thereto on agricultural products and foodstuffs. Brussels, 21.12.2005, COM(2005) 671 final, 2005/0278 (CNS), 2005/0279 (CNS), http://europa.eu.int/eurlex/ lex/LexUriServ/site/en/com/2005/com2005_0671en01.pdf (accessed 2008-05-02)

FAO (2003) Proceedings of 2003 International Organic Conferences, Bangkok, Thailand

FAO/WHO (2001) Guidelines for the Production, Processing, Labeling and Marketing of Organically Produced Foods; GL 32-1999, Rev.1-2001. Joint FAO/WHO Food Standards Programme, Rome, 2001

Green Net / Earth Net Foundation (2005) Organic Agriculture Situation in Thailand 2005 (Thai language), mimeograph, Earth Net Foundation, Bangkok

International Trade Centre (ITC) (2006) The European Market for Organic Fruit and Vegetables from Thailand. Strengthening the Export Capacity of Thailand's Organic Agriculture- Final Report

Kramol, P.; Thong-ngam, K.; Gypmantasiri, P. & Davies, W.P. (2006) Challenges in Developing Pesticide-free and Organic Vegetable Markets and Farming Systems for Smallholder Farmers in North Thailand. Proc. Ith IS on Supply Chains in Transitional Econ., Ed.: P.J. Batt. Acta Hort. **699**, ISHS 2006

National Office of Agricultural Commodity and Food Standards, Ministry of Agriculture and Cooperatives. (2003) Organic Agriculture, Part 1: The Production, Processing, Labelling and Marketing of Organic Produce. ACFS, Bangkok

NESDB (1997) The 8th National Economic and Social Development Plan 1997-2001

Office of Organic Agriculture Standards (MGT) (2003) Organic Agriculture Standards. Ministry of Agriculture and Cooperatives

Panyakul, V. (2002) Introduction to Organic Agriculture, Earth Net Foundation, Bangkok

Panyakul, V. & Sukajirattikal, J. (2003) Situation of Organic Agriculture in Thailand and the World, Earth Net Foundation, Bangkok

Panyakul, V. & Sukajirattikal, J. (2003) Markets of Organic Products, Earth Net Foundation, Bangkok

Panyakul, V. (2004) Organic Agriculture: How to Get Certification, Earth Net Foundation, Bangkok

UN-ESCAP (2002) Organic Agriculture and Rural Poverty Alleviation: Potential and Best Practices in Asia. Poverty Reduction Section, UN-ESCAP / Poverty and Development Division (PDD), UNESCAP, Bangkok

Unno, A. (2003) Sustainable Agriculture Movement in Thai Society. The Sustainable Agriculture Foundation Thailand. Bangkok, Thailand. (in Thai)

USDA (2006) Thailand Organic Products: Thailand's Organic Outlook 2006. GAIN
Report No: TH6007, 19/1/2006. Foreign Agricultural Service, Bangkok

Willer, H. & Yussefi, M. (eds.) (2004) The World of Organic Agriculture: Statistics and Emerging Trends 2004. International Federation of Organic Agriculture Movements, Bonn

Organic food and human nutrition

Claus Leitzmann, Isabel Behrendt

Institute of Nutrition, Justus Liebig University, Giessen, Germany
E-Mail: ClausLeitzmann@aol.com

8.1 Introduction and definitions

Never doubt that a small group of people could change the world.
Indeed, it is the only thing that ever has.
Margaret Mead 1901-1978

Organic food is defined as food produced by Organic or Biological Agriculture and usually certified as such. These descriptive terms can be confusing, since agriculture of all types is in fundamental senses both organic and biological. Organic food is produced by specified and certified low- or no-chemical input methods usually in countries whose main agriculture has become industrialized. It should be recognized that such agriculture is the traditional agriculture, still practiced in parts of the world where farmers cannot afford or choose not to use chemical inputs.

By convention the term *Organic Agriculture* is used to describe various agricultural systems that have developed since the 1920s. These are distinctly different both from traditional indigenous subsistence farming still used in many middle- and low-income countries, and high input industrialised agriculture, as now practiced in most parts of the world.

The definition of Organic Agriculture by the Federation of Organic Agricultural Movements (IFOAM) is based on a set of general principles whose aim is to achieve an ideal agricultural system. Organic Agriculture is defined as "a whole system approach based upon a set of processes resulting in a sustainable ecosystem, safe food, good nutrition, animal welfare and social justice" (IFOAM 2002). The Codex Alimentarius Commission (2007) emphasizes that certified Organic Agriculture is a holistic production management system, implying that it is an integral part of the complete ecosystem. Organic food is defined as a product of this holistic ecosystem (Leitzmann 2002).

Demand for organic food is growing rapidly. Farmland dedicated to organic production increased fourfold within the last decade. Although Organic Farming originated in Germany in the 1920s before it spread to other European countries and North America, the current expansion of Organic Farming is particularly rapid in tropical and subtropical countries, which now contain about 75% of the worldwide area with certified Organic Farming. However, presently more than 90% of the certified organic food produced worldwide is consumed in Europe and the USA (Halberg et al 2006). This means that people in the tropics are likely to be generally unaware of the potential benefits of reliably certified and advertised organic food to health.

It has to be pointed out that many farmers in middle- and low-income countries produce *uncertified* organic food, because they do not have the money for either agricultural chemicals or for certification. In many countries there is a big internal market for *uncertified* organic food.

Nutrient-dense, minimally contaminated organic food is likely not only to improve physical health but also mental and emotional health and general well-being. Consumers of organic food are also citizens who by their choice play an active part to maintain an intact environment and contribute to a better world (Köpke 2005).

8.2 Organic Agriculture in the tropics

The challenges faced by tropical organic agricultural systems are quite different from those in temperate zones. It is not clear whether principles developed for Organic Farming in high-income countries usually in the Northern hemisphere, apply to middle- and low-income countries in the tropics and subtropics, and whether these can or even should be adapted to traditional knowledge and local conditions.

Organic Agriculture in the tropics and subtropics currently produces mainly plant foods, because there are very limited local markets for organic foods of animal origin. However, in Organic Agriculture animal and plant production should always be combined to ensure a closed cycle of material flow and to enhance food diversity (Abele 2006).
Also, currently Organic Agriculture in the tropics and subtropics apparently produces mainly cash crops for export; but lack of information about organic food production for direct consumption and for local markets is almost certainly underestimated.

Organic food production in the tropics and subtropics is an increasing source of income for families and small scale farmers, farmers' co-operatives and larger commercial farms. But there are barriers to entry in the Western markets, such as the lack of internationally recognised certification bodies and limited experience in production and marketing. There is also growing competition from different regions and the possibility of market saturation. In tropical and subtropical countries, especially in Africa and Latin America, trade in certified organic produce is mostly orientated towards foreign markets. The main reason for this is the great need of cash to pay international debts.

One suggestion to develop local markets for organic food is to employ the method that has been successfully used for the introduction of other foods and drinks: the imitation of the eating habits of health-conscious people consuming organic food (Johannson 2005).

In addition to families and other small farmers, co-operatives can develop and strengthen domestic markets for certified organic food. Problems include weak local infrastructures, poorly developed national markets, high prices, high costs for certification, and inadequate national laws. Since the distances between producer and consumer are usually short (direct sale or sale on domestic markets), certification seems to be unnecessary for both farmer and customer. Some of the main issues with organic foods are whether they are added value for human health, why they cost more, whether they help to alleviate poverty and hunger in tropical countries, and whether they contribute to national and international food and nutrition security.

This article mainly deals with the potential health benefits of organic food (Lundegardh & Martensson 2003; Cooper et al. 2007). The other questions will be touched upon, but for details see other chapters of this book (see also Leitzmann et al 2004-2008).

Most of the research on the health value of organic food was carried out in temperate zones. The pertinent scientific data will be presented and is meant to stimulate discussion and research by scientists of different backgrounds working in tropical and subtropical countries.

8.3 Human nutrition

Tell me what you eat and I will tell you what you are.
Jean Anthelme Brillat-Savarin, 1755-1826

A short review of some basic aspects of human nutrition as related to organic food will be given. Further details are found in a number of current textbooks (Elmadfa & Leitzmann 2004; Geissler & Powers 2005; Gibney et al 2002; Mann & Truswell 2007; Shils et al. 2006).

All life is nourished by the nutrients contained in foods. These provide energy and the building materials that are essential for growth, health and reproduction of living beings. Foods differ greatly worldwide and are eaten for a variety of reasons. Once inside the alimentary tract, digestion reduces food to nutrients, which determine the quality of the nourishment.

Basically, the requirements for nutrients of the human body are rather similar worldwide, but are influenced quantitatively and qualitatively by genetic heritage, gender, stature, physical activity and health status. The average nutrient requirements of the same person living in temperate or in tropical regions are similar. Personal dietary energy needs can be substantially different, depending among other things on gender, age, body weight and physical activity. In tropical regions a higher intake of fluids is required.

8.3.1 Nutrition in the life cycle

Healthy old age starts early in life.
German proverb

Nutrition has to be adapted to specific times of growth, development and aging. Maternal and possibly paternal nutrition prior to conception may affect the health of the newborn. Both proper nutrition during pregnancy and exclusive breastfeeding for six months benefit the health of the infant at the time and later in life, and also the mother throughout life including in post-childbearing years.

Good food habits and sustained physical activity during childhood lessen the possibility of inappropriate eating behaviour, which now happens with disturbing frequency during adolescence. The general effect of nutrition on morbidity and mortality usually becomes manifest in later adulthood. Therefore, the prevention of chronic diseases that tend to appear later in life should begin in childhood.

Pregnant and nursing women have some extra energy and nutrient requirements, but these are less than usually believed, since hormonal changes will increase the use of nutrients at these times. The energy needs of adults gradually decrease in later middle and old age, while some nutrients like vitamin D and B_{12} are required in higher amounts.

Since the second half of the 20th century malnutrition (in the sense of inadequate nutrition) has mainly been a major public health issue with children in middle- and low-income countries, usually in the tropics and subtropics. Elderly people in these parts of the world may also be malnourished. Care for the elderly has traditionally been provided in the family, but this culture is beginning to disappear. AIDS is particularly devastating in tropical and subtropical countries (Ntozi & Zirimenya 1999).

8.3.2 Nutrition for health and wellbeing

An apple a day keeps the doctor away.
Pembrokeshire proverb 1866

The relation of nutrition and disease has long been established. It is increasingly clear that the incidence of various diseases like cancer, arteriosclerotic heart disease, hypertension, diabetes, and also overweight and obesity, can be largely prevented by practicing appropriate nutrition. Generally these are still relatively uncommon in rural regions, in the tropics, but are now common in cities throughout the world, as people abandon their traditional ways of life and behave and eat like people in high-income countries.

Prevention of chronic non-communicable diseases is not only an individual responsibility. With the rapid increase of non-infectious diseases, overweight and obesity and associated health problems worldwide, governments need to provide essential infrastructure and financial support. Infrastructure includes among many others ensuring wholesome food systems and safe water, a health care system with professionally trained staff, and well equipped hospitals. Many tropical countries lack these basic essentials.

Sustained physical activity enhances health and well-being, helps to prevent many disorders and diseases, and is the most effective way to control body weight. Once people become very overweight attempts to reduce excess body fat by dieting are rarely successful. The best time to guard against overweight and also chronic non-communicable diseases, which are now becoming increasingly common in tropical countries, is early in life.

8.3.3 Food safety

It is the dosage that makes a toxin poisonous.
Paracelsus (1493-1541)

Governments and the public are right to be concerned about food safety. Consumers worry most about genetic engineering of food, food additives, food radiation and the residues from agricultural use of pesticides, antibiotics and hormones (see chapter 8.7).

Microbial contamination of food and water can also be dangerous, especially to children and other vulnerable people, and indeed can be deadly. The presence of natural toxicants in foods is an important public health issue in many tropical and subtropical countries. Pathogenic microbes are often present in undercooked meat, especially poultry, dairy products and eggs. Transmission also takes place through improper food handling in the food system or at home. In hot damp tropical countries prepared dishes that are kept unrefrigerated for an extended period of time are likely to cause enteric diseases. Thorough cooking, proper food hygiene and refrigeration of leftovers will help to avoid these problems.

Some substances occurring naturally in foods can be deadly, as is well known of some mushrooms. Other substances are toxic, like solanin in the green parts of potatoes. Aflatoxin, a well known carcinogen of high potency, is a product of a mould that occurs mainly in peanuts and cereal grains, particularly in hot damp climates.

8.3.4 Nutritional care in disease

Food, the best medicine.
Old proverb

Nutrition-related diseases can be prevented, or at least delayed, by maintaining appropriate nutrition and regular physical activity throughout life. These include disorders and diseases associated with chronic undernutrition, which are common in many tropical and sub-tropical countries particularly in Asia and Africa, and also diseases of which the cause is acute starvation, which remains common mostly in some African and Asian countries. Undernutrition usually implies shortage of most if not all essential nutrients including dietary energy, amino acids, essential fats, vitamins, minerals and other bioactive substances. Those now usually identified as most common in undernourishment include shortage or deficiency of iron, zinc, iodine, vitamin A and folate. Combined deficiency of dietary energy and of protein is also identified as endemic especially in some parts of sub-Saharan Africa and South Asia. Diseases of specific nutrients are best prevented or cured by proper nutrition.

Until the second half of the 20th century chronic non-communicable diseases were public health issues only in high-income countries and indeed were once termed 'diseases of civilisation' or 'Western diseases'. But with mechanisation and urbanisation and increasing availability of money, these diseases are now epidemic in middle- and even low-income countries in the tropics and subtropics. Some of the highest rates of overweight and obesity are now found in such countries. This is because of the replacement of traditional by industrial diets, and reduction of physical activity (Prentice 2006).

8.4 Common foods for human nutrition

The fate of nations is determined by what they eat.
Jean Anthelme Brillat-Savarin, 1755-1826

While nutrients are what nourish the body, it is not nutrients but food that is grown or purchased, prepared, and consumed. Foods for human nutrition can be grouped into categories with common characteristics (Table 1).

Table 1: Common foods and some of their characteristics

Cereals (wheat, rice, maize; also sorghum, barley, oats, rye)

- Usually staple foods, contain a ratio of macronutrients (protein: carbohydrates : fats) that is close to human requirements
- Whole grain cereals and their products contain essential minerals (K, Ca, Mg, Fe, Zn), a large number of vitamins (E, B_1, B_2, B_6, niacin, pantothenic acid, folic acid, biotin), high amounts of dietary fibre and different kinds of secondary plant substances
- Practically devoid of vitamin C and (except for maize) of beta-carotene (a precursor of vitamin A)
- In many tropical regions cereals are staple foods, especially rice, maize and sorghum
- Buckwheat (*Fagopyrum esculentum*), quinoa (*Chenopodium quinoa)* and amaranth (*Amaranthus edulis)* do not belong to the cereal family, but they contain comparable kinds and amounts of nutrients. Quinoa and amaranth are mainly grown in tropical areas

Tubers and roots (potatoes, plantains, sweet potatoes, yams, cassava (manioc, yucca, also taro)

- Contain mainly complex carbohydrates and serve in many tropical regions as staple foods
- Often poor sources of protein. Contain frequently minerals (Mg, Fe, K) and vitamins (C, thiamine, niacin)
- Cheap source of dietary energy and a good source of dietary fibre

Legumes (soybeans, peanuts, beans, peas, lentils; also chickpeas, lupines, sweet peas)

- Contain high concentrations of protein and dietary fibre and a fair amount of complex carbohydrates, vitamins (thiamine, niacin, pantothenic acid), minerals (Mg, P, K, Fe), and other bioactive substances
- Low fat content, except for soybeans and peanuts
- Grown and consumed mainly in tropical and subtropical countries; excellent source of dietary energy

Vegetables

- Contain the highest concentrations of most essential minerals, vitamins, other bioactive substances, and soluble dietary fibre
- Deliver a fair amount of complex carbohydrates, some protein and very little fat
- Green leafy vegetables are good sources of iron, riboflavin, folate and beta-carotene. Different colours of vegetables indicate the presence of different bioactive substances
- Large differences in nutrient concentrations are common in the many existing varieties of vegetables
- Many vegetables originated in tropical regions where they are still grown and consumed in basic meals

Fruits and berries

- Rich source of many vitamins with anti-oxidative properties like vitamin C
- Contain numerous bioactive substances and soluble dietary fibre
- Fat is present only in small amounts, except for avocados and coconuts
- Many are grown and consumed in tropical regions
- Export is a sizeable part of the cash income in the tropics

Table 1: Common foods and some of their characteristics (cont)

Oil seeds (soybeans, rape, sunflower, palm; also linseed, hemp, cotton, poppy, pumpkin, sesame, grape, pine) - Contain high amounts of various oils - Contain fair amounts of protein, except for soybeans with high concentrations - The native plant oils contain high concentrations of vitamin E and the B-vitamins and some minerals (Mg, P, Zn, Co, Mn) - Many oil seeds and oil fruits are grown primarily in tropical regions
Nuts (hazelnut, walnut, cashew, almond; also Brazil nut, macadamia, pistachio, chestnut) - Contain similar nutrients as oil seeds - Ideal as snacks - Grown mostly in tropical and subtropical regions
Herbs and spices - Contain many bioactive substances, some with antimicrobial properties - Many mainly grown in tropical regions
Animal products (meat, milk, eggs; also marine animals) - Contain a number of valuable nutrients like protein, iron, zinc and vitamin B_{12} - The production and market structure for certified organic foods of animal origin is generally poorly developed in tropical and subtropical countries

Many of these foods are native to tropical regions and are best grown – or can only be grown – in tropical climates. Nutritional adequacy is best achieved by consuming foods from all the groups identified here, however, diets that are plant-based, or which exclude some or all animal foods, can also be nutritionally adequate.

8.5 The foundation of organic food

Organic food is the product of a broad-based food system connected with a number of different fields of knowledge that are touched upon in the following paragraphs. Some of these fields are crucial in tropical regions. Detailed descriptions and examples are given in other chapters of this series.

Nutrition ecology is an interdisciplinary scientific discipline referring to complete food systems that encompass Organic Agriculture (Leitzmann 2003). It embraces the practice of agricultural production (use of machinery, energy, fertilizers and pesticides) and all of the components that follow throughout food systems, including harvesting, preservation and storage over transport, processing, packing and trade to distribution, preparation and consumption of food, including the disposal of all waste materials.

Nutrition ecology also encompasses interactions of food systems with personal and public health, and our economy, society and living and physical environment. A principle of nutrition ecology is food systems that are indefinitely sustainable. The diets deriving from these systems are predominantly plant derived, originate from Organic Agriculture, are produced mostly locally and regionally, are minimally processed, ecologically packaged, fair traded and tastefully prepared. Such diets are also socially acceptable, economically feasible, and culturally desirable. (Koerber et al 2004).

Sustainability should be a vital concern for all agricultural systems, especially in tropical regions where food security is crucial. Soils are often poor or degraded and climate can be erratic. Nutrition ecology is a concept also concerned with human health and well-being (Leitzmann 2003a and b).

The *New Nutrition Science* project is a recent initiative whose purpose is to give the science of nutrition a broader definition. As an integrated science it encompasses biological, social and environmental dimensions all of which are permeated by economic realities. The classic biological dimension of nutrition will remain central, concerned with the interactions of food and nutrition with physiological, metabolic and now also genomic systems, and the effects of these interactions with health and disease.

The project evaluates food systems, foods and drinks, and their nutrients and other constituents within and between all relevant biological, social, environmental and also economic systems. It is proposed that the linked political, financial and electronic revolutions known as 'globalisation', together with scientific and biotechnical discoveries, and demographic, epidemiological and nutritional trends, themselves imply a new conceptual framework for nutrition, in order to analyse and assess all relevant determinants of well-being, health and disease, and to take effective action.

The New Nutrition Science project follows ethical and ecological principles, recognises history, tradition and culture, and affirms human rights. It is committed to the development and maintenance of policies and programmes designed to preserve and maintain the human, living and physical world (Leitzmann & Cannon 2005).

Eco-nutrition includes biodiversity or food diversity and means variety of different traditional crops that are part of many eco-systems, with the potential to produce organic food. This additional source of nutrients is especially important in low-income countries, where decreased food variety and monocultures and the substitution of food crops by cash crops are widespread. For example, in the past millet was the main food produced and consumed in parts of Asia and Africa which is now gradually becoming substituted by maize, which has a higher productivity than millet. Currently maize is either exported or used as bio-fuel. The world food and energy market and industrialisation supersede traditional and usually sensible eating habits in tropical countries. Eco-nutrition describes a nutrition pattern based on intact eco-systems (Wahlquist & Specht 1998; Wahlquist 2005).

Indigenous foods are often organic foods by definition, since in many cases they are cultivated in traditional farming systems by farmers who do not have the money for chemical inputs. They expand food variety and help to stabilise food and nutrition security, because they are often available free or at a low cost, and contain various essential nutrients and help to avoid nutrient deficiencies still common in some tropical regions. Many indigenous wild foods are valued more by pharmaceutical companies as sources of drugs, and are neglected as regional and local food sources. Sometimes this is because knowledge of their value as food is overlooked or lost.

Intercropping is an agricultural practice of cultivating two or more crops in the same field at the same time. This system is considered as well-suited for Organic Agriculture in the tropics. It increases biodiversity and food and nutrition security. The various products ensure self-supply and food for sale on local markets. In Cuba, for example, beans, maize, vegetables and avocados are produced within citrus cultures. Some farmers also keep animals like sheep, chicken or pigs in this intercropping area (Kilcher 2005).

Aquaculture is a potential means to produce organic foods. It is defined as the farming of aquatic animals and plants in the same field, like the combination of fish in rice fields, which is also another possibility to increase food diversity. The main factor governing a greater consumption of fish in tropical countries is the lower price and therefore greater affordability compared with other animal protein sources (Scialabba et al 2003).

8.6 Nutrient content of organic food

Not the rich, the sensible buy organic food
Recent cognition

Plants need a variety of nutrients to display their full genetic potential. The deficiency of any essential nutrient will limit plant growth. Based on current scientific judgement, conventional agriculture supplies all the known nutrients necessary for optimal growth for a particular crop with a mixture of inorganic mineral fertilisers. The prerequisites for this strategy are expert knowledge and sufficient money, both of which are often not available in tropical and subtropical countries.

Inappropriate use of mineral fertilisers degrades food quality. As for fossil energy, the global supply of some mineral fertilisers is limited and will not be available in the future. Currently conventional industrialised agricultural systems are unlikely to be sustainable.

In Organic Agriculture the ideal nutritional mix is supplied with crop residues, green manure, leguminous plants, solid dung and liquid manures. This method is based on traditional knowledge as well as an appropriate crop rotation design and requires minimal cash.

The question whether organically produced foods are more nutritious is of central interest. Information remains somewhat limited, since most data available originate from studies that analysed either organic *or* conventionally produced food, sometimes grown under different soil and climate conditions and even in different years, so that any conclusions drawn from these comparisons are unsure. Further research under standardised conditions is needed.

In general, the influence of organic food on human health aspects is assumed to be positive, since a number of nutrients are found in higher concentrations. Animal experiments show that organic food improves reproductive function, and resistance to infections. When given a choice, animals most often prefer organic feed to conventional rations (Velimirov 2001).

The quantity of nutrients synthesised by the plant does not only depend on the method of farming. This is determined largely by the genetic make-up of the cultivated plant variety and is influenced by soil quality, climatic condition and in some cases the time of harvest. The available database is not broad, and more research is needed to substantiate the preliminary conclusions drawn here.

The following sections present comparisons of the nutrient contents of foods produced by either conventional or Organic Agriculture.

8.6.1 Protein

The amount of protein needed to maintain good health and resistance to disease for humans is usually overestimated. The daily average requirement for adults is somewhat less than 1g/kg body weight. Almost all food patterns in adequate quantity normally supply this amount of protein. In high-income countries typically more than ample protein is consumed. In low-income countries whose staple foods include tubers and roots that are poor sources of protein, such as cassava or sweet potatoes, overall protein consumption can be deficient, and this remains a public health issue especially when these are given in the form of porridges as weaning foods. In general, protein deficiency in the tropics is of lesser concern today as food supplied become more varied and when there is access to sufficient affordable quantities of food. Poverty and ignorance remain basic causes of protein deficiency.

Like all other substances, the protein content of plants is genetically determined. However the final quantity and quality can be influenced to a certain extent by the availability of nitrogen. Increased nitrogen uptake will increase the protein content. Beyond the optimal storage level, proteins of lesser nutritional quality are synthesised due to lower amounts of essential amino acids.

Another noticeable effect of excessive nitrogen uptake is decreased fragrance and aroma of foods. This is due to certain secondary metabolites (Table 2).

The quality of proteins in foods – also called their biological value – depends on the relationship of their amino acid patterns to the human need of amino acids. On this basis single plant proteins usually have a lower protein quality than foods of animal origin, due to the lower ratio of essential or indispensable to nonessential or dispensable amino acids. However, in mixed meals of plant foods, the protein quality is improved by complementation of the different amino acid patterns. For example, legumes may have a high content of the essential amino acids lysine and tryptophan, both of which are low in cereals. Cereals are high in methionine, which is the limiting amino acid in legumes. A combination of these stable foods (for example wheat, rice or maize, plus beans or other legumes), produces a protein quality as high as that of meat.

The traditional cuisines of many South American countries is characterised by maize (and later rice) and beans. They are either consumed alone, or in combination with other foods, usually when of animal origin eaten only sparingly. Such dietary patterns may be primarily related to the foods available in these countries and not to the awareness of their beneficial health effects.

Organic food usually contains less protein than industrially produced food, but the protein is of higher quality. Protein quality is influenced by nitrogen from any kind of fertilisers. The excessive nitrogen uptake leads in addition to increased protein also to decreased carbohydrate synthesis (Worthington 2001).

8.6.2 Minerals

Minerals in human nutrition are classified by the amounts present in the human body. Minerals that are found in quantities of more than 50mg/kg body weight are called macro-minerals, whereas micro-minerals are found in amounts below this level, and are further subdivided into trace and ultra-trace elements. About half of the 2-3 kg of body minerals is composed of calcium, with the other macro-minerals (sodium, phosphorus, magnesium, chloride, potassium, sulphur) constituting most of the other half.

A small amount is made up of the trace elements (in decreasing amounts: iron, zinc, copper, selenium, manganese, iodine, molybdenum, nickel, cobalt) and ultra-trace elements (in decreasing amounts: silicon, bromine, lead, strontium, aluminium, boron, tin, cadmium, germanium, arsenic, mercury, chromium, rubidium, lithium, vanadium).

The genetic make-up of the plant determines the efficiency with which minerals are taken up from the soil or from the different fertilisers applied. The availability of minerals is also determined by the presence of organic material, which leads among others to a symbiosis with soil fungi and an increased microbial activity in the soil, both of which will greatly enhance the effective root surface.

Most studies show that the contents of mineral and trace elements (Ca, Fe, P, Mn, Zn, Cu und K) in foods produced by industrial and Organic Agriculture show no significant differences (Magkos et al 2003, Williams 2002, Williamson 2007, Woese et al 1997, Worthington 2001).

8.6.3 Vitamins

Fat-soluble vitamins (A, D, E and K) are found mainly in foods of animal origin. Particularly vitamin A is a critical nutrient in many of the tropical regions. Among others, vitamin A is essential for the proper functioning of the eye, deficiencies lead to a number of eye sight problems and eventually to blindness. Millions of children, especially in tropical and subtropical countries, go blind every year due to vitamin A deficiency. Beta-carotene found in plants is a precursor of vitamin A. One study showed no differences in vitamin A or beta-carotene contents in vegetables between the two groups (Williams 2002).

Vitamin D is essential for proper bone development and deficiencies cause delayed bone growth and osteoporosis. Vitamin D can be synthesized from precursors in the skin with the aid of sunlight. Good sources of vitamin D are fish, meat and mushrooms as well as fortified milk products. There are some comparative data on the quantities of fat-soluble vitamins in conventional and organic foods, showing slight differences (Bergamo et al 2003).

Water-soluble vitamins are numerous: thiamine [B_1], riboflavin [B_2], pantothenic acid [B_3], niacin [B_5], pyridoxine [B_6], biotin [B_7], folate [B_9], cobalamin [B_{12}], and vitamin C. Some of these are critical for vulnerable populations in the tropics and subtropics. Thiamine and riboflavin deficiencies are most often observed. High-income vegetarian populations tend to have vitamin B_{12} deficiencies, whereas vegetarians living in unhygienic conditions do not show this deficiency. This is because micro-organisms on hands and food and in water are sources of vitamin B_{12}.

A comparison of the vitamin concentrations in organic foods compared with conventional foods shows no measurable differences in vitamin B levels for cereals, potatoes and vegetables.

For vitamin C there is good evidence that higher amounts are found in organically grown potatoes and vegetables, due to an increased availability of nitrogen. As is the case for all nutrients, vitamin C concentrations in fruits and vegetables are influenced by various factors such as genetic make-up, soil and climate conditions, cultural practice, maturity as well as time and methods of harvesting (Alföldi 2001, Magkos et al 2003, Tauscher et al 2003, Williams 2002).

8.6.4 Dietary fibre

The term dietary fibre refers to a number of plant cell wall substances that chemically are non-starch polysaccharides. The most common of these is cellulose followed by hemicelluloses, both of which are found in wholegrain wheat, rye, barley and oats. Grains most often consumed in tropical and subtropical regions like milled rice and maize are rather low in dietary fibre. The characteristic dietary fibre of fruits is pectin.

In tropical and subtropical countries fruits are commonly only seasonally consumed and then often not in large quantities, so that the dietary fibre intake from fruits is usually low. Legumes, often consumed as a staple food throughout the year, contain large amounts of dietary fibre.

Organic foods usually contain larger amounts of dietary fibre, since the spacing of plants is different from conventional systems (plants are grown in less dense stands) and receive less nitrogen and are harvested closely to their optimal time of ripeness.

8.6.5 Other bioactive substances

Plants contain a wide range of bioactive substances also called secondary metabolites or phytochemicals. In recent decades there has been an increasing recognition of their health benefits. Phytochemicals are partly responsible for the protective effects of vegetables and fruits, nuts, cereals and legumes against various cancers and cardiovascular disease. In addition they have antioxidative, immunmodulatory, antimicrobial and cholesterol-lowering properties. Other health promoting effects of phytochemicals include regulation of blood pressure, blood glucose levels, blood coagulation and inhibition of inflammatory processes. In sum, phytochemicals are important determinants of the health value of foods (Watzl & Leitzmann 2005).

Of all substances investigated so far, the differences are largest between the content of phytochemicals in organic food compared with industrially produced foods. Organically grown vegetables have been considered as to contain between 10-50% more phytochemicals compared to conventionally grown plants (Brandt & Molgard 2001; Cooper et al. 2007) because the plants are given more space to develop and less nitrogen is available for plant uptake. This allows a shortened vegetative growth period, and an earlier onset of maturation during which more phytochemicals are synthesised. Also the crops are harvested as close to their optimal time of ripeness as possible (Table 2).

Table 2: Increased phytochemical content of organic foods

Phytochemical	Food	Literature
Beta-carotene	carrot	Tauscher et al 2003; Velimirow et al 2003
Lycopene	tomato	Tauscher et al 2003; Velimirow et al 2003
Polyphenols	potato	Tauscher et al 2003; Velimirow et al 2003
Polyphenols	pears, peaches	Carbonaro et al 2002
Resveratrol	red wine	Levite et al 2001; Velimirow et al 2003
Flavonol	apple	Weibel et al 2001
Sulforaphane	broccoli	Adams 2001

In addition to these increases in the amount of individual phytochemicals higher antimutagenic effects (Ren et al 2001) and an increased antioxidative potential (Ebata et al 1993) were determined in organically grown leafy vegetables. A new report based on the findings of 97 studies finds that the nutrient content, especially antioxidants, in organic food is markedly higher when compared to conventionally produced foods (Benbrook et al, 2008).

8.7 Residues and other harmful substances

Food quality depends also on presence or absence of harmful chemical substances. Since the application of pesticides and other chemical substances and the use of antibiotics and hormones are not permitted in Organic Agriculture, the residues and heavy metals detected in organic food are found only in very small amounts. Contamination can occur by acid rain, from neighbouring fields, or by errors in processing, packaging and distribution.

The main contaminating substances found in food show the influence of industrialisation (Table 3).

Table 3: Residues and other harmful substances in foods

Pesticide residues and their metabolites are contained in much lower amounts in organically grown food than industrially grown products. The pesticides found in organic food are either due to driftage from other fields, old environmental damage or insufficient separation during transport, storage, processing or sale. Sometimes illegitimate use of pesticides has been evident. Alföldi et al (2006) reported organic foods containing a medium pesticide level of 0.002 mg/kg fruit and about 0.003 mg/kg vegetable. In comparison, the medium pesticide content of industrially grown vegetables and fruits was 0.4 mg/ kg

Table 3: Residues and other harmful substances in foods (cont)

Contaminants found in foods like lead, cadmium and mercury are independent of the method of agriculture. These metals originate from emissions of traffic and industry and accumulate in the soil (Alföldi et al 2006).
Nitrates were found in lower amounts in organic food like leafy vegetables and amounted to a range of 10-40% of nitrate content determined in industrially grown foods. Normally the maximum residual levels in organic food are not exceeded (Alföldi et al 2006).
Antibiotics and *hormones* are not allowed in Organic Agriculture for raising animals, so that they are not to be found. In any event, in low income tropical and subtropical countries animal derived foods are generally less important.

Food quality tends to decline when agriculture is rapidly industrialised. China, where the extent of cropland per head is rather small, is an example. This limitation is compensated with increased crop output by additional manpower and the use of high amounts of fertilisers and pesticides. The current goals of the government to attain rapid economic growth and self-sufficiency with food, further increases the excessive use of fertilisers and pesticides. Results include water shortage and degradation of cultivated land, which threaten food safety and food security. Now there is some recognition in China of the health and environmental benefits of Organic Agriculture (Zhao et al 2007).

8.8 Food preservation, processing and preparation

Many facts are known,
however by different persons.
Werner Kollath, 1892-1970

Handling of crops after harvest has an influence on the food quality. Various methods of preservation, processing and preparation increase the shelf-life but usually decrease the nutritional value of foods. In certain tropical areas some of these methods have long traditions, whereas others are less known or even unknown.

8.8.1 Food preservation

The various traditional and industrial methods of food preservation influence the nutritional quality of food. How food is preserved is just as important as how it is produced. Some traditional food preservation methods have been developed in tropical regions, since foods tend to spoil much quicker in hot and humid climates (Table 4).

Table 4: Methods of food preservation

Fermentation - Means of increasing the nutrient content of food which has been practised for millennia - Refrigeration reduces the use fermentation, but in tropical countries, fermentation is sometimes one of the most used methods of food preservation - Many different plants, especially of the cabbage (Brassica) family, but also foods of animal origin like fish, are traditionally conserved by fermentation

Table 4: Methods of food preservation (cont)

Drying - Time-tested method used in almost all parts of the world - Natural practice that leads to losses of some vitamins and other bioactive plant substances
Smoking - Natural way to preserve foods with some losses of the vulnerable ingredients Some of the many constituents of smoke can cause cancer
Salting - Used for millennia to preserve plant and animal derived foods with little losses in the activity of nutrients - With the introduction of refrigerators this method has been reserved for fish and some traditional foods like sauerkraut - Decline of salting has been followed by a reduction in the incidence of stomach cancer
Freezing - Natural method used for millennia in areas with long and cold winters - Excellent means of preserving the nutrients of foods - Used in households and food industry but the proper infrastructure (electricity, appliances) are still sometimes not available in tropical regions
Irradiation - Used to kill micro-organisms to increase the shelf-life of foods - Probably harmless in its other effects

8.8.2 Food processing

Foods are processed for preservation and extension of shelf-life, to remove inedible parts, to increase digestibility and the bioavailability of some nutrients, to improve palatability and texture, to eliminate micro-organisms, to destroy toxins and antinutritional factors. Some food products are totally processed – made up of processed ingredients plus additives. Some of the different methods of food processing have been developed to a very sophisticated level (Table 5).

Table 5: Different methods of food processing

Milling - Leads to substantial losses of nutrients and dietary fibres amounting up to 90% for milled grains - Chronic non-communicable diseases were uncommon in Africa until well into the second half of the 20th century, partly because of high intakes of dietary fibre contained in unrefined foods
Pasteurisation / Sterilisation - Heat treatment of foods at about 75°C to kill pathogenic micro-organisms - Heat treatment of food at 121°C to kill all micro-organisms and their spores - Loss of nutrients depends on the duration of treatment
Canning - Causes degradation of some flavour and some vitamins over time
Extrusion - Homogenisation method often integrated in heat processes - Breaks up fat globules and other compounded substances into smaller fragments, which changes their absorption rate

Up to 90% of industrially produced food supplied in high-income countries is processed by these and other methods. The percentages in tropical and subtropical countries rise with increasing use of money and in many urban areas are much the same as in high-income countries

8.8.3 Food preparation

Heating is the most frequently used method to prepare food. All methods of thermal processing improve the digestibility of proteins and carbohydrates. To what extent protein is denatured by heating is still being debated, since the stomach acids are sufficient for that purpose. Complex carbohydrates in potatoes have to be gelatinised by heating before they can be digested, otherwise they can cause problems in the large intestine. Heating can increase the bioavailability of nutrients and secondary plant substances like beta-carotene in carrots and lycopene in tomatoes.

Minerals are not damaged by heating, but will be leached into the cooking water, which can itself be consumed or used as a base for soups and sauces. The vitamins, other bioactive substances and dietary fibres are reduced in their activity by heating. These losses vary. Vitamin C is the most sensitive vitamin to heat exposure and is used as an indicator of heat damage to nutrients.

Additional means of heating of food other than boiling and steaming include stewing, baking, grilling/broiling, roasting, frying and barbecuing. Most of these food preparation methods cause higher nutrient losses and may lead to the production of cancer-causing substances.

During microwave cooking, heat is generated within the food-product by microwave-induced molecular vibrations. The oscillation creates intermolecular friction that generates heat. The nutrient losses can be as high as or higher than with conventional cooking methods.

8.9 Summary

Make sustainable choices, easy choices.
Recent cognition

Certified organic food is a product of whole-system agriculture. It provides safe nourishing food. It also protects animal welfare and social justice. Associated with the concepts of Organic Agriculture and food are the concepts of nutrition ecology and eco-nutrition, which encourage good agricultural and food practices and protect biodiversity and indigenous plants and propagate intercropping and aquaculture.

The nutritional requirements of humans differ during the life cycle but are fairly independent of geographical location. Many common foods are native to tropical and subtropical regions. Health benefits of organic food are likely to derive from higher contents of some bioactive substances, also known as phytochemicals. These have high anti-mutagenic effects and an increased antioxidative potential. Use of pesticides and other chemicals are not permitted in Organic Agriculture, so residues and heavy metals are rarely found in organic food.

Methods of food preservation, processing and preparation affect organic foods in the same ways as industrially produced foods.

References

Abele, S.; Dubois, T.; Twine, E.; Sonder, K. & Coulibaly, O. (2006) Organic agriculture in Africa: a critical review from a multidisciplinary perspective. J Agric Rural Develop Trop Subtrop, 143-166. In: Hülsebusch, C.; Wichern, F.; Hemann, H. & Wolff, P. (eds.) Organic agriculture in the tropics and subtropics – current status and perspectives, Beiheft 89

Adam S. (2001) Vergleich des Gehaltes an Glucoraphanin in Broccoli aus konventionellem und aus ökologischem Anbau. Bundesanstalt für Ernährung (Hrsg.) Jahresbericht 2001

Alföldi, T.; Bickel, R. & Weibel, F. (2001) Vergleichende Qualitätsuntersuchung – neue Ansätze und Impulse täten gut. Ökologie& Landbau **117**, 11-13

Alföldi, T.; Granado, J.; Kieffer, E.; Kretzschmar, U.; Morgner, M.; Niggli, U.; Schädeli, A.; Speiser, B.; Weibel, F.; Schmidt, G. & Schmidt; W. (2006) Qualität und Sicherheit von Bioprodukten. FIBL (eds.) Lebensmittel im Vergleich (Dossier Nr. 4)

Benbrook, C.; Zhao, X.; Yáñez, J.; Davies, N. & Andrews, P. (2008) State of Science Review: Nutritional Superiority of Organic Foods. http://www.organic-center.org/reportfiles/5367_Nutrient_Content_SSR_FINAL_V2.pdf (accessed 2008-04-25)

Cannon, G. & Leitzmann, C. (2006) New science for the new millennium (editorial). Asia Pac J Clin Nutr **50** (1), 2-5

Cannon, G. & Leitzmann, C. (2006) The New Nutrition Science project. Scand J Food Nutr **50** (1), 5-12

Carbonaro, M.; Mattera, M.; Nicoli, S; Bergamo, P. & Cappelloni, M. (2002) Modulation of antioxidant compounds in organic vs conventional fruit (peach, prune and pear) J Agric Food Chem **50**, 5458-5462

Codex Alimentarius Commission (2007) Guidelines for the production, labelling and marketing of organically produced foods. FAO, Rome

Cooper, J.; Niggli, U. & Leifert, C. (2007) Handbook of organic food safety and quality. Woodhead Publishing, Cambridge, England

Ebata, J.; Kawai, K. & Furukawa, H. (1993) Inhibitory effect of dietars leafy vegetables on mutagens and on active oxygens. Brozetti, G. et al (eds.) Antimutagenesis and anticarcinogenesis mechanisms III. Plenum Press, New York, 99-102

Elmadfa, I. & Leitzmann, C. (2004) Ernährung des Menschen. Ulmer Verlag, Stuttgart

Geissler, C. & Powers, H. (eds) (2005) Human nutrition. Churchill Livingstone, Edinburgh

Gibney, M., Vorster, H. & Kok, F. (2002) Introduction to human nutrition. Blackwell Publishing, Oxford

Halberg, N.; Alroe, H.F.; Knudsen, M.T. & Kristensen, E.S. (eds.) (2006) Global development of organic agriculture: challenges and prospects. CABI Publishing, Wallingford/UK

Holt, G. & Reed, M. (eds.) (2006) Sociological perspectives of organic agriculture: from pioneer to policy. CABI Publishing, Wallingford/UK

IFOAM (2002) International Federation of Organic Agriculture Movements. Basic standards for organic production. Tholex-Theley, Germany

Johannsen, J.; Mertineit, A.; Wilhelm, B.; Buntzel-Cano, R.; Schöne, F. & Fleckenstein, M. (2005) Ökologische Landwirtschaft. Ein Beitrag zur nachhaltigen Armutsbekämpfung in Entwicklungsländern? Forum Umwelt und Entwicklung, Bonn. Internet: http://www.eed.de/fix/files/doc/eed_Forum_oekologische_Landwirtschaft_05_deu.pdf (accessed 2008-04-28)

Kilcher, L. (2005) Biolandbau als Chance für eine nachhaltige Entwicklung. Ökologie & Landbau **135** (3), 14-17

Köpke, U. (2005) Organic foods: do they have a role? In: Elmadfa, I. Diet Diversifikation and health promotion. Forum Nutr **57**, 62-72

Koerber, K. v.; Männle, T. & Leitzmann, C. (2004) Vollwert-Ernährung: Konzeption einer zeitgemäßen und nachhaltigen Ernährungsweise. Haug Verlag, Stuttgart

Kristiansen, P.; Taji, A. & Reganold, J. (eds.) (2006) Organic farming - a global perspective. CSIRO Publishing, Collingwood/ Australia

Leitzmann, C. (2001) Die globale Ernährungssituation - Welternährung zu Beginn des 21. Jahrhunderts. Biologie unserer Zeit **31**, 408-416

Leitzmann, C. (2002) Was ist ein Ökoprodukt bzw. ein Bio-Lebensmittel? – Erwartungen und Tatsachen. Zeitschrift für das gesamte Lebensmittelrecht **29**, 151-167

Leitzmann, C. (2003a) Nutrition ecology. Elmadfa, I. (ed) Diet diversification and health promotion. Forum Nutr **56**, 220-221

Leitzmann, C. (2003 b) Nutrition ecology: the contribution of vegetarian diets. Am J Clin Nutr **78**, 697-699

Leitzmann, C.; Beck, A.; Hamm, U. & Hermanowski, R. (2004-2008) Handbuch Bio-Lebensmittel, Loseblattsammlung. Behr's Verlag, Hamburg

Leitzmann, C. & Cannon, G. (2005) The New Nutrition Science project. Publ Health Nutr **8**, 667-804

Levite, D., Adrian, M. & Tamm, L. (2000) Preliminary results of resveratrol in wine of organic and conventional vineyards. Proc 6th Internat Congr Org Viticulture, 257-258

Lundegardh, B. & Martensson, A. (2003) Organically produced plant food – evidence of health benefits. Acta Agric Scand, Section B – Plant Soil Science **53** (1), 3-15

Magkos, F.; Arvaniti, F. & Zampelas, A. (2001) Organic food: nutritious food or food for thought? A review of evidence. Int J Food Sci Nutr **54** (5), 357-371

Mann, J. & Truswell, S. (2007) Essentials of human nutrition. Oxford University Press, Oxford

Ntozi, J.P.M. & Zirimenya S (1999) Changes in household composition and family structure during the AIDS epidemic in Uganda. Continuing Afr HIV/AIDS Epidemic **1**, 193-209

Prentice, A.M. (2006) The emerging epidemic of obesity in developing countries. Int J Epidemiol **35**, 93-99

Ren, H.; Endo, H. & Hayashi, T. (2001) Antioxidative and mutagenic activities and polyphenol content of pesticide-free and organically cultivated green vegetables using water-soluble chitosan as a soil modifier and leaf surface spray. Sci Food Agric **81**, 1426-1432

Scialabba N. & Hattam C (2003) Organic agriculture, environment and food security. Environment and natural resources series no. 4. Internet: http://www.fao.org/DOCREP/005/Y4137E/Y4137E00.HTM (accessed 2007-11-30)

Shils, M. E.; Olson, J. A.; Sheike, M. & Ross, A. C. (2006) Modern nutrition in health and disease. Williams & Wilkins, Baltimore

Tauscher, B.; Brack, G.; Flachowsky, G.; Henning, M.; Köpke, U.; Meier-Ploeger, A.; Münzing, K.; Niggli, U.; Pabst, K.; Rahmann, G.; Willhöft, C. & Mayer-Miebach, E. (2003) Bewertung von Lebensmittel verschiedener Produktionsverfahren, Statusbericht 2003. Internet: http://www.bmelv.de/cln_044/nn_749122/SharedDocs/ downloads/03Ernaehrung/LebensmittelVergleich,templateId=raw,property =publicationFile.pdf/LebensmittelVergleich.pdf (accessed 2007-11-27)

Velimirow, A. (2001) Ratten bevorzugen Biofutter. Ökologie & Landbau **117** (1), 19-21

Velimirow, A. & Müller, W. (2003) Die Qualität biologisch erzeugter Lebensmittel. Ergebnis einer umfassenden Literaturrecherche. Wien

Wahlquist, M.L. & Specht, R.L. (1998) Food variety and biodiversity: Econutrition Asia Pacific J Clin Nutr **7** (3/4), 314-319

Wahlquist , M.L. (2005) Diversification in indigenous and ethnic food culture. In: Elmadfa, I. (ed): Diet diversification and health promotion. Forum Nutr. **57**, 52-61

Watzl, B. & Leitzmann, C. (2005) Bioaktive Substanzen in Lebensmitteln. Hippokrates Verlag, Stuttgart

Weibel, F.P.; Bickel, R.; Leuthold, S.; Alföldi, T.; Niggli, U.; & Balzer-Graf, U. (2001) Bioäpfel – besser und gesünder? Eine Vergleichsstudie mit Standard- und Alternativmethoden der Qualitätserfassung. Ökologie & Landbau **117** (1), 25-28

Williams, C.M. (2002) Nutritional quality of organic food: shades of grey or shades of green? Proc Nutr Soc **61**, 19-24

Woese, K.; Lange, D.; Boess C. & Werner Bogl, K. (1997) A comparison of organically and conventionally grown foods – results of a review of the relevant literature. J Science Food Agric **74**, 281-293

Worthington, V. (2001) Nutritional quality of organic versus conventional fruits, vegetables and grains. J Altern Complement Med **2**, 161-173

Zhao, J.; Luo, Q.; Deng, H. & Yan, Y. (2007) Opportunities and challenges of sustainable agriculture development in China. Phil Trans R Soc B **31** (in press)

Organic Agriculture in the Mediterranean: Ten years of cooperation and networking with CIHEAM - IAMB

Lina Al Bitar; Maurizio Raeli

Mediterranean Agronomic Institute, Via Ceglie, 9, 70010 Valenzano (Bari), Italy
E-Mail: albitar@iamb.it

9.1 Introduction

The Mediterranean Agronomic Institute of Bari (IAMB) is the Italian offshoot of the International Centre for Advanced Mediterranean Agronomic Studies (CIHEAM), an international inter-governmental organisation established in 1962 under the patronage of the OECD and the Council of Europe, which presently groups 13 Mediterranean countries including Albania, Algeria, Egypt, France, Greece, Italy, Lebanon, Malta, Morocco, Portugal, Spain, Tunisia and Turkey.

CIHEAM is situated in Paris where the General Secretariat is headquartered and promotes and co-ordinates the activities of the four Institutes located in Bari (Italy), Chania (Greece), Montpellier (France) and Saragoza (Spain).

CIHEAM in general and the Institute of Bari in particular closely follow the evolution of agriculture in the Mediterranean countries and the impact of new technologies and cultivation techniques on economy, farm management and land use.

Since 1997, IAMB has endeavoured to become a landmark for the Organic Agriculture sector by providing training, research and experimentation and promoting Networking to co-operate to the development of the Mediterranean basin, thus duplicating its previous achievements in "Land and water resources management: irrigated agriculture" and "Integrated pest management of Mediterranean fruit crops".

Over the last years, Organic Agriculture has attracted much attention in the Mediterranean countries for both environmental reasons and market opportunities and by now it involves nearly all Mediterranean countries (Al Bitar, 2004).

However, despite its progressive development and the rising interest, the organic sector in the Mediterranean region is still constrained by many critical points and is drawing heavily on the Northern European regulatory, technical and scientific model.

In a report issued by UNCTAD Secretary (United Nations Conference on Trade and Development - July 1996), the United Nations stated that Organic Agriculture could make a concrete contribution to development while improving environmental conditions, raising people's income and ensuring food security also in developing countries.

This was also pointed out in the conclusions of the first Euro-Mediterranean ministerial conference on agriculture, which took place in Venice on the 26th of November 2003. The conclusions identified Organic Agriculture as one of the main factors of the economic development for the countries in the Mediterranean region (Lacirignola, 2004).

9.2 Historical background

In 1997, IAMB started its activities in the field of Organic Agriculture. The initial drive came from the European Union, which included some measures promoting and supporting organic production in the Multi-fund Operational Programmes (MOP). IAMB developed a project entitled ' Biopuglia Project: demonstrative action to develop and spread the organic production method in agriculture' sponsored by the Apulian regional government within the Apulia MOP 1994-99. This project aimed at providing public institutions, associations, professionals and farmers of Apulia with a telematic access via the Internet to updated information concerning agronomic production techniques, industrial processing, the legal framework of the sector and international market trends.

At the same time, the 7th AgriBioMediterraneo Conference was held in Bari, July 21-23, 1997 and officially announced the establishment of an IFOAM Mediterranean Regional Group with its secretariat based at IAMB.

The purpose of the initiative was to tackle specific and structural problems confronting Mediterranean Organic Agriculture, especially in research and experimentation, education and training.

IAMB has firmly committed itself to coordinate the group by fostering the exchange of technical expertise to help the development of the region and harmonising the actions and goals of the various Mediterranean countries and institutions to design a common action policy catering for the interests and needs of all the regional stakeholders.

Organic Agriculture has much scope for progress in the Mediterranean basin, especially in those areas where planning an industrial and intensive agriculture model to compete with worldwide production would be a mere illusion.

Against this background, IAMB has committed itself to foster and sustain agricultural development in the Mediterranean basin by accomplishing the following:

- The provision of technical and scientific training and support to technicians, academics and trainers of in-depth knowledge of Organic Agriculture;
- The launching of a programme of research, investigation and experimentation focusing on the main Mediterranean crops under Organic Farming;
- The establishment of co-operation programmes and active participation in the establishment of a network of Organic Agriculture in the Mediterranean area.

9.3 The mission

Training, Research and Cooperation became the missions of IAMB also in the field of Organic Agriculture. Through these 3 missions, IAMB contributes to developing a shared Mediterranean organic identity in its member countries ,and a common approach facing the big food and agriculture challenges of tomorrow with one objective: to let this approach be recognized at the international level.

9.3.1 Training to excellence

IAMB guideline is matching theoretical learning with the ability to apply knowledge acquired to the solution of real local problems.

Therefore, training programmes aim at providing both a basic scientific and technical qualified knowledge and an appropriate training to the use of technologies that can reasonably be applied in the countries of origin of trainees, where they will eventually go back to operate.
Admission is granted to well selected young graduates, researchers, ministry officials or university teachers and researchers coming from Mediterranean and near East Countries. Moreover, a limited number of trainees, selected by international organisations of co-operation to development, come from Asia, Africa and South America to attend the courses.

There are two types of training at IAMB: the long-term and the short-term courses.

9.3.1.1 The long-term or post-graduate training

This training programme started in 2000 and is organised over two years:

- A one-year course to obtain the Post-graduate Diploma (DSPU) runs over 30 weeks including 700 hours of theoretical and practical work. Specialisation methods and contents are complementary to training programmes run at national level in the CIHEAM member states. This training programme follows a multidisciplinary approach; it has an international and Mediterranean focus and favours a cultural and linguistic exchange. Teaching is ensured by lecturers-researchers from all countries, either permanent faculty members or eminent visiting lecturers, chosen for their expertise and professional experience.

The diversity of the teaching staff ensures a better understanding of the subject and provides opportunities for co-operation between representatives of the countries involved in the different programmes. More particularly, it helps promote the "*spirit of international cooperation*" that the founders of the Centre wished to bring about.

The teaching system complies with a European standard that adopts a Credit Transfer System (ECTS) designed to enable students to accumulate transferable credits that will facilitate comparability and equivalence and promote student mobility and access to work right across Europe.

- A two-year course to obtain the diploma of *Master of Science* in which students, who have successfully accomplished the DSPU and have met all the pre-requisites set by the Institute, have to complete an additional individual research project on one of the aspects of Organic Agriculture carried out over one year (12 months). This activity could be conducted in a residential form in Bari or in a mobility form in one of the CIHEAM member countries under an International co-supervision.

The Master programme in Mediterranean Organic Agriculture is designed to train candidates for future professional careers by developing skills related to production and management; legislation, inspection, certification and labeling; socio-economic analysis and market strategy and to provide them with the necessary tools and expertise to assess the agricultural, environmental, and socio-economic opportunities and constraints of Organic Agriculture in different Mediterranean areas. It also allows participants to learn the approach to research in Organic Farming as well as to prepare them for scientific research especially for those willing to apply for the PhD programmes carried out jointly between IAMB and several Italian Universities.

Over 200 students have followed the post-graduate training at IAMB yet. Table 1 illustrates the total number of students divided by country and by typology of the training.

Table 1: Number and origin of students who have followed post-graduate training

Country of origin	DSPU	MASTER			PhD
		At IAM-B	Mobility	Total MASTER	
ALBANIA	7	2	1	3	
ALGERIA	8	3	0	3	
BANGLADESH	1	1	0	1	
BOSNIA & HERZEGOVINA	5	2	0	2	
CROATIA	1	0	0	0	
EGYPT	12	4	5	9	
JORDAN	5	2	1	3	
ITALY	1	0	0	0	4
LEBANON	13	10	1	11	5
MACEDONIA	1	1	0	1	
MONTENEGRO	1	1	0	1	
MOROCCO	15	6	7	13	2
PALESTINE	10	1	1	2	
PORTUGAL	1	0	0	0	
P.R. OF CHINA	3	1	0	1	
SERBIA	4	1	0	1	1
SYRIA	10	2	1	3	
SPAIN	1	0	0	0	
TUNISIA	16	2	9	11	
TURKEY	11	3	5	8	
TOTAL	126	42	31	73	12
GENERAL TOTAL			211		

It is noteworthy to mention that quite all the former trainees embraced the path of research and teaching and many have occupied a prestigious position as directors or department executives in many local or foreign institutions operating in the field of Organic Agriculture.

9.3.1.2 The short-term training

These are national or international short cycles of few weeks addressed to professionals of public and private sectors, teachers, researchers or extensionists wishing to expand and update their knowledge of Organic Agriculture methods. They are carried out at IAMB or in the member countries.

These cycles are usually planned in partnership with institutions of the member countries in both curriculum development and the implementation of the training session.
Since 1998 up to today, 36 courses have been organized at IAMB or in the member countries (Table 2) where 884 students coming from 34 countries were taught (Table 3) in all fields of Organic Agriculture.

Table 2: Specialised Intensive Short Courses on Organic Agriculture

Course title	Country	Period	Year
Organic Agriculture: legislation, agronomic techniques and marketing	IAM Bari - Italy	September 26 - October 11	1998
Production Methods in Organic Agriculture: Normative Principles and Technical Aspects	Izmir - Turkey	June 14-23	1999
Inspection and Certification of Organic Production in the Mediterranean Region	IAM-Bari	September 5-19	1999
Cours Intensif en Agriculture Biologique: Principes et pratiques	Tunis - Tunisia	November 29 - December 8	1999
Techniques de Production des Principales Cultures Méditerranéennes	Sousse - Tunisia	Novembre 19-26	2000
Gestion Intégrée d'une ferme biologique	Sousse - Tunisia	November 5-14	2001
Organic Vegetable Production in the Mediterranean Basin	Cairo - Egypt	March 20-31	2002
Diplôme de Spécialisation en agriculture biologique *Première session* "Principes et pratiques"	Sousse -Tunisia	December 10-20	2002
Training for e-learning tutors in Organic Farming	IAM-Bari	May 16-17	2003
Inspection and Certification of Organic Production Outside EU	IAM-Bari	September 4-20	2003
Manager dei processi di innovazione della filiera Agroalimentare Bio	IAM-Bari	September 2002	December 2003
Tecnologie e processi innovativi per la tracciabilità, la verifica e il controllo della qualità nelle produzioni biologiche ortofrutticole	IAM-Bari	March 2003	December 2004
Organic and Fair trade inspection	IAM-Bari	October 4-15	2004

Table 2 (cont)

Course title	Country	Period	Year
Session de perfectionnement et de recyclage en Agriculture biologique: arboriculture fruitière	Sousse - Tunisia	December 12-18	2004
Setting up and implementation of sustainable and multifunctional rural development model based on organic and competitive agriculture: the use of SEBINA platform	IAM-BARI	February	2004
Setting up and implementation of sustainable and multifunctional rural development model based on organic and competitive agriculture	Chania- Greece	September	2004
Researching Rural Development & Organic Farming in the EU-25 context	IAM-Bari	September 27–October 2	2004
first course on Inspection and certification procedures in Organic Farming	Tirana	October 3-8	2005
Session de perfectionnement et de recyclage en agriculture biologique "Grandes Cultures et élevage"	Sousse - Tunisia	December 11-18	2005
Session de perfectionnement et de recyclage en agriculture biologique "Arboricolture fruitère"	Tunisia	December	2005
Summer Training	IAM-BARI	July-September	2002
Summer Training	IAM-BARI	July-September	2003
Summer Training	IAM-BARI	July-September	2004
Summer Training	IAM-BARI	July-September	2005
Summer Training	IAM-BARI	July-September	2006
Summer Training	IAM-BARI	July-September	2007
second course on Inspection and certification procedures in Organic Farming	Tirana	March 27-31	2006
Manager in processi innovativi di compostaggio e gestione delle biomasse di scarto	IAM-Bari	May 2005 - July 2006	
Organic Agriculture: Principles and technical Aspects	IAM-Bari	July 17-27	2006
Upgrading on Organic and Fair-trade Certification	IAM-Bari	October 18-22	2006
Strengthening of services to farms and institutional support for the development of Organic Agriculture	IAM-Bari	October 22-29	2006
Expérience méditerranéenne en: Maraîchage Biologique	Sousse-Tunisie	December 11-16	2006
Organic Agriculture Training Program	IAM-Bari	July 3 - August 31	2007
Establishing common models of integrated sustainable monitoring, planning and management of high environmental value areas to control natural resources degradation	IAM-Bari	July 4-6	2007

Table 2 (cont)

1st course on Training of Technical experts in Organic Agriculture, in support of rural development and of food emergency in the Balkan area	IAM-Bari	February 5 - March 4	2007
2nd course on Training of Technical experts in Organic Agriculture, in support of rural development and of food emergency in the Balkan area	IAM-Bari	July 9-22	2007

Table 3: Number of students who have followed the short term training by country of origin and gender distribution

COUNTRY	Gender division		Total
	M	F	
Albania	34	17	51
Algeria	16	5	21
Brazil	2		2
Cile	1		1
Croatia	6	11	17
Cyprus	1		1
Czech Republic	1		1
Egypt	13	2	15
Bosnia & Herzegovina	3	7	10
France	1		1
Germany	1	1	2
Greece	10	4	14
Italy	100	69	169
Israel	1	1	2
Jordan		1	1
Lebanon	11	18	29
Macedonia	1	2	3
Malta	4	1	5
Montenegro	3	3	6
Morocco	10	2	12
Palestine	5		5
Poland	3	1	4
Portugal	1	1	2
Romania		3	3
Saudi Arabia	5		5
Serbia	3	3	6
Slovakia		2	2
Slovenia	1	1	2
Spain		1	1
Syria	1		1
The Netherlands	1		1
Tunisia	333	80	413
Turkey	40	36	76
TOTAL	612	272	884

9.3.2 Research

In order to cater for research and experimentation requirements in Organic Agriculture, IAMB launched in 1988 some activities to tackle the whole host of significant problems confronting Mediterranean organic productions.

IAMB invests with its partners in view of building common references, criteria of definition and an approach mainly aiming at the resolution of problems related to Organic Agriculture in a Mediterranean context.

IAMB takes part in or directly manages various research programmes, particularly those secured under the tendering procedure for the EU's Research Framework Programme. Research is carried out most often as part of a co-operative exercise involving researchers and experts from different Mediterranean countries. In this way it is possible to avoid duplication of work in particular areas and to encourage synergies.

IAMB promotes comparative research in different Mediterranean countries through its Master mobility programme and its Mediterranean Organic Agriculture Network (MOAN).

Research activities have focused on the following topical issues:

9.3.2.1 Agroecology

This activity focuses on the description of the system in all its components to make a diagnosis and give useful indications for its management and for land use planning.
This area covers cropping system management, sustainability analysis, energy balance of agricultural production and biodiversity protection, enhancement and preservation.

9.3.2.2 Soil fertility management and composting

The main objective is the proper management of soil fertility through the use of the farm residues with special reference to quality compost.
This area focuses mainly on comparing different types of compost, optimizing the production techniques, assessing and optimizing the compost tea production and application and assessing their impact on disease management.

9.3.2.3 Pest and disease management

The objective is to contribute to the improvement of the global quality of Mediterranean organic products, mainly olive and grape, through the validation of pest management methods and products by studying their efficacy and degradability, the rearing, releasing and genetic characterisation of some beneficial insects (mainly *Psyttalia concolor*) and the isolation and identification of new biological control agents against grapevine powdery mildew.

9.3.2.4 Socio-economic and market studies

This area has a twofold objective that consists of exploring and assessing the development potential and prospects of Organic Agriculture in the Mediterranean region and analysing the design and implementation of co-operation and training initiatives to further promote the sector. It mainly covers farm performance and supply chain analysis, policies and agro-food marketing and the assessment of the organic sector's impact on poverty alleviation and rural development.

9.3.3 International Cooperation to development

Over the last years, IAMB has intensified its on-the-spot activities within the framework of international cooperation programmes funded by the Italian Ministry of Foreign Affairs, whose Directorate General-Cooperation for Development is IAMB basic counterpart, the FAO, the World Bank, the European Union, GTZ, OADA and other international and national organisations involved in cooperation for development. Emphasis is laid on the study and implementation of projects within Interreg programmes, Multifund Action Plans, and Multiregional Action Plans funded by the European Union.

Through these cooperation agreements and programmes, IAMB promotes quite significant partnership actions in the Mediterranean region contributing to upgrade and develop the agricultural sector and protect the environment with a view to progressively spread sustainable and Organic Agriculture through the following actions:

- Enhancing human resources as an all-important factor of sustainable development in Mediterranean agriculture;
- Deepening, widening and spreading scientific and technological knowledge as a basic factor to improve production techniques;
- Disseminating the culture of international co-operation as a key-factor to socio-economic development and peaceful coexistence of Mediterranean people.

In particular, the IAMB's distinctive function is to act as a stimulus and a catalyst, proposing initiatives that go along with *in-loco* actions that involve local stakeholders – governmental and scientific institutions, technical and enterprise communities – for them to collaborate closely with each other during the study and implementation stages, thus favouring utmost synergism.

The scientific institutions of the involved countries are invited to participate in the design of the proposed actions relying upon the know-how and experience of IAMB, upon its specific skills, facilities and equipment, as well as its human resources.

In view of the European Union enlargement eastwards, IAMB has also expanded local partnership projects in the Balkan area. At the request of some international organisations (FAO, IFAD), these activities are being extended to countries bordering the Near-East (Iraq and Iran) where problems in agriculture are similar to those faced by the Mediterranean regions.

Hereinafter are listed some of the cooperation projects that IAMB has promoted to contribute to territorial development of the beneficiary countries.

9.3.3.1 Institutional Building for Organic Agriculture in Syria

It is a three-year (2006-2008) project, promoted by IAMB and financed by the Italian Ministry for Foreign Affairs through the FAO-Italian Government Cooperative Programme. The beneficiary institution is the General Commission for Scientific Agricultural Research -Syrian Ministry of Agriculture.

The project will facilitate the realization of a platform for the sustainable development of Organic Farming in Syria, through the elaboration of a proper institutional framework, made by the following components:

a) elaboration of a national legislation,
b) elaboration of guidelines for the establishment of a certification body,
c) information, training and education of an adequate number of Syrian nationals (leader farmers, technicians, scientists and opinion leaders),
d) market studies to ensure long-term market driven economic sustainability of the proposed activities and
e) an applied research program will be defined and partially implemented.

Within the frame of this project, IAMB has been appointed to organise and coordinate training activities to be carried out in Italy and in Syria, the organisations of study tours and training sessions for the Ministry staff and the organic unit, general support and technical backstopping of the training activities and the technical contribution and support for the compilation of didactic material.

9.3.3.2 Training of Technical experts in Organic Agriculture, in support of rural development and of food emergency in the Balkan area (Bio- 84)

This one year project (December 2006- December 2007) was promoted by IAMB and financed by the Italian Ministry of Foreign Affairs within the Law 84/01 Balkans – provision for the Italian participation in stabilisation, reconstruction and development of the countries of the Balkan area.

The main objective was to favour the development of Organic Agriculture in the beneficiary countries through the implementation of specific training actions to the personnel of the local institutions, institutional strengthening, support to research and networking.

The partner institutions are the Ministries of Agriculture of Albania, Bosnia & Herzegovina, Croatia, Macedonia, Montenegro and Serbia.

9.3.3.3 Strengthening of services to farms and institutional support for the development of Organic Agriculture (SIAB)

The project is financed by the European Commission Adriatic Cross Border INTERREG IIIA, Axis 3 – Actions to strengthen cooperation; Measure 3.1 Human resources training, professional refreshing and innovative initiatives in the field of social promotion and the job market; Action 3.1.1 – Cross-border professional training (September 2005 – August 2007).

The main objective is the development of quality farming through qualification and strengthening of sector institutions and organisations in 5 countries (Albania, Bosnia & Herzegovina, Croatia, Montenegro and Serbia) through the:

- Improvement of cross border cooperation in the primary sector.
- Institutional strengthening and harmonisation of systems: creation of a normative and institutional context through professional qualification and training in the public administration and in the institutions entrusted with the management of the Organic Farming system.

- Qualification of professional skills and strengthening of services offered to farms to convert to the organic production method.
- Dissemination of know-how.

9.3.3.4 PAB – INTERREG IIIA Italy-Albania

Financed by the European Union through the European Fund for Regional Development Interreg IIIA 2000-2006. It has a duration of 3 years and involves the Apulia Region (Italy) and Albania.

It consists mainly in strengthening the scientific partnership between Italy and Albania in the field of Organic Agriculture with the aim to enhance the know-how and the access to scientific information. For this aim, a scientific network including Italian and Albanian scientists was formed. The activities included: validation of technical standards of the main crops of economic interest for the area; amelioration of services and technical units; establishment of a competence centre; the realisation of a model for technical assistance based on the realisation of technical support to farms on-line; institutional capacity building; technical know-how exchange; professional skills development towards a national action plan.

The project gave birth to the first Albanian national strategy to promote the organic sector. Further developments towards the elaboration of a National Action Plans have been envisaged in a further extension of the project.

9.3.3.5 Supporting activities to the Tunisian fruit and vegetable productions

Financed by the Italian Ministry for Foreign Affairs in the framework of International cooperation programmes and co-financed by IAMB. This project has a duration of three years (2007-2010). One of the main activities is the institutional backstopping to the Tunisian institutions, mainly the Ministry of Agriculture. It consists of the regulatory framework updating, training of technicians in Italy, guidelines elaboration for organic grapevine and artichoke production which are early crops and very much requested by the international market. In fact, some partnerships have also been created between Tunisian farmers and Italian entrepreneurs.

9.3.3.6 Setting up and implementation of sustainable and multifunctional rural development model based on Organic and Competitive Agriculture - SIMOCA"

INTERREG III B (2000-2006) CADSES – Central European Adriatic Danubian South-Eastern European Space.

The project (January 2003 - March 2006), implemented by IAMB, involved many Italian public institutions, and other academic institutions from Greece, Croatia, Poland and Slovakia.

The main objective was to define a new strategy for a sustainable and multi-functional rural development based on Organic Agriculture. For this reason, the territorial integration among the CADSES countries was favoured by the exchange of know-how among the different organisations dealing with the promotion and the development of the territory (decision makers, final users etc.) through the constitution of international networks.

The project has favoured the following:

1. Building up of a common transnational knowledge shared in a permanent communication system.
2. Elaboration of a model of best practices of an organic farm in the selected areas.
3. Setting up and validation of a transnational strategy of sustainable, multifunctional rural development.
4. Training of competences for the dissemination of innovative systems.
5. The creation of different competence centres in the project countries, linked altogether by a polycentric documentation centre.

9.3.3.7 Activation of Mechanisms to Sustain Rural Territories & Communities in Lebanon – TERCOM.

The main objective of the project starting in March 2007 and ending in March 2008 is the identification of three areas of intervention for the support of the rural communities damaged by the last war of July 2006.

One of these areas is Organic Agriculture, where the project aims at assisting the Lebanese government in the following actions:

1. The revision and the improvement of the draft national law on Organic Agriculture before its submission to the Parliament for the final approval.
2. Re-structuring, organisation and staffing of the ministerial unit dealing with Organic Agriculture and on-the-job training of the ministerial officers on all the issues related to their tasks.

9.4 Networking

Besides the cooperation projects, in 1999 IAMB launched a network of experts called MOAN (Mediterranean Organic Agriculture Network) with the objective of further developing Organic Agriculture in the Mediterranean countries.

MOAN gathered institutions from 13 countries beside IFOAM Agribiomediterraneo, and realised a series of initiatives of promotion, training, research and data collection on the state-of-the-art of the organic sector in the Mediterranean basin.

In 2006, a new organisation of MOAN was planned that foresees its enlargement to 24 Euro-Mediterranean countries and the active participation of the Ministry of Agriculture of each member country, in particular the unit in charge of Organic Agriculture, with the aim to promote the organic sector in both the EU and non-member Mediterranean countries and to strengthen the Euro-Mediterranean partnership.

The three objectives of MOAN are the following:

1. **Creating a homogenous partnership** that would be able to discuss all the problems related to the institutional organisation of the sector at regional level and look for common strategies for its development.

2. **Strengthening the relations** among organic communities (especially among official institutions like Ministries) in the Mediterranean region.
3. **Valorisation of the potential** of Organic Agriculture in the region.

An important step to meet the challenges listed above is the support of **information and knowledge sharing** among national and international organic communities through the:

- organisation of an annual meeting with all the network members; if necessary separate meetings on different topics might be organised, to discuss problems related to the sector and exchange and share common strategies;
- organisation of country-specific workshops which could encourage the creation of national sub-networks, enhance the visibility of the host country and raise its profile in the Mediterranean organic context;
- organisation of international conferences and publication of scientific works, periodical reports on the state-of-the-art of OA in the region, popular works etc. with the aim to **bring the Mediterranean organic perspective in the global debate;**
- creation and management of a website on Mediterranean OA with special sections on country profiles and events.

Table 4: MOAN member countries

European countries	*Eastern Adriatic countries*	*Eastern and Southern Mediterranean countries*
• Cyprus*	• Albania	• Algeria
• France	• Bosnia&Herzegovina	• Egypt
• Greece*	• Croatia	• Jordan
• Italy	• Macedonia	• Lebanon
• Malta	• Serbia	• Libya
• Portugal*	• Montenegro	• Morocco
• Spain		• Palestine
• Slovenia		• Syria
		• Tunisia
		• Turkey

* have not joined officially yet

MOAN (http://moan.iamb.it) was launched in November 2006 in Bari. The first meeting was attended by 16 representatives of Mediterranean countries, out of the 24 invited.

A number of specific objectives were fixed for this first meeting:

- Create a common and shared knowledge of the state-of-the-art of Organic Agriculture
- Exchange best practice experiences
- Lay the foundation for encouraging the emergence of a Mediterranean organic identity through the sharing of common problems and needs
- Generate ideas on future collective initiatives in the framework of MOAN
- Prioritise MOAN specific objectives and future thematic focuses

According to the results of the priority setting exercise it appeared that there is scope for MOAN to play an important role in representing and promoting Mediterranean identity and specificities in Organic Agriculture on the international arena.

In relation to that, reliable and detailed information about the sector, its strengths and weaknesses as well as an objective assessment of Organic Agriculture's multifaceted potential and impacts on Mediterranean farming systems and rural communities is quintessential. Therefore, data collection, analysis & dissemination and impact evaluation represent key activities for MOAN in the near future.

At the same time, participants also emphasised the need for comparative research and institutional cooperation work aimed at the building up of a shared knowledge of organic regulatory and policy issues and the design and implementation of capacity building and institutional strengthening initiatives on such issues across the Mediterranean.

The second MOAN meeting was held in June 2007 in Turkey and focused mainly on the "Methodological issues in data collection on Organic Agriculture in the Mediterranean". Nineteen countries participated.

Four main objectives were fixed for the meeting and connected to follow-up activities:

- Review the current state-of-the-art of organic data collection and reporting in MOAN countries
- Exchange best practices in organic data collection, analysis and dissemination
- Identify common problems and challenges related to organic data collection and reporting
- Launch a long-term programme of activities and initiatives for the development of a shared framework for the collection and reporting of reliable data on the Mediterranean organic sector.

The meeting ended with the official constitution of the National Turkish Network, the first national sub-network established in the framework of MOAN. Similar initiatives will be encouraged in all the member countries hosting the next MOAN meetings.

From 16 to 20 June 2008, IFOAM will hold its 16th Organic World Congress "Cultivate the Future" in Modena (IT), including ISOFAR's 2nd International Scientific Conference. The MOAN will primarily contribute to the Congress through the organisation of a specific event on the theme "Public support policies for the development of Organic Agriculture in the Mediterranean" to which all MOAN member countries will be invited to participate and to actively contribute.

References

Al-Bitar, L. (Ed.) (2004) Report on Organic Agriculture in the Mediterranean Area – Mediterranean Organic Agriculture Network, Options Méditerranéennes, Series B: N°50, CIHEAM- IAMB, Bari

Lacirignola, C. (2004) Eight years after Barcelona dialogue has begun. Greenmed, the Mediterranean fresh produce magazine **1**, 26-31

Case Study

Somalia: Phytopathological situation and organic plant protection perspectives

Giorgio M. Balestra

Dipartimento di Protezione delle Piante, Università degli Studi della Tuscia, Via S. Camillo de Lellis, 01100, Viterbo, Italy. E-mail: balestra@unitus.it

10.0 Abstract

Considering the persistent unstable political situation, the Somalian agriculture is one of the most disadvantaged of the world and in particular among those of African countries. Thanks to the UNA NGO's *Consortium*, in 2001 a project (still in progress and supported by the EU) started to promote the Integrated Pest Management (IPM) principles in Somalia as part of a major international effort to assist in the reconstruction of this country after 17 years of civil war. Here are synthesised 5 years of field investigations related to the main Somalian phytopathological problems. Plant pathogens (bacteria, fungi virus) have been recorded on the most important horticultural plants, cereals and stone fruits (cucurbits, eggplant, onion, pepper, tomato, maize, sorghum, rice, cotton, sugar cane, sesame, sunflower, banana, citrus, mango, date palm and papaya) crops in Somalia. Moreover, natural strategies and first organic plant protection temptatives were developed to spread the fundamental bases of an agricultural able to preserve this environment and to give a safe and real opportunity to rebuild the future of Somali farmers.

Key words: Africa, biocontrol, IPM, natural extracts, climate warming, phytopathogenic bacteria.

10.1 The activities

The project was developed in Somalia by three phases. Phase I of the EU project (promotion of IPM principles in Somalia) started off by recognizing that the small quantities of agrochemicals sold in the country were being used in the most irresponsible manner and several of the few products present were also highly toxic. Top priority was given to implement some safe use trainings. Initial training started in the State of Puntland in 2001, through the production of simple safety messages newssheets that were widely circulated in the region. Following this activity, UNA NGO's consortium decided that the safe use module should receive further priority in attention. The former booklet on "Safe Use of Pesticides" used in Kenya was upgraded, translated into Somali and printed in large numbers for countrywide distribution. At the same time a teaching manual "Training Formats for Safe Use of Pesticides" was developed and again printed in a large number of copies. The books have formed the basis of all training work in Somalia these last years and both items have recently been reprinted. A second phase of the project, approved in 2003 and again on chemical safe use, was a key component. Much of this training was to be directed at the farmer community level. During the early part of this phase a set of high quality posters was developed depicting typical safe use messages which were produced for use at the farmer field days. Many training sessions were held countrywide, training hundreds of stockists, extensionists and students as well as farmers. Most of these sessions were held on the field, in the villages or under trees. As phase II was reaching its end, three IPM Training and Promotion Centres in different regions were set up (Garowe, the capital of Puntland State, Hargeisa in Somaliland and Jowhar in Central-South Somalia). These centres, when fully operational, will offer IPM trainings and visibility work in their respective regions. They have all been provided with modern communication equipment, sprayers and protective clothes for demonstration purposes.

In 2006, hard work in conjunction with the FAO was rewarded with a third phase IPM contract (Consolidation of the Integrated Pest Management Initiative in Somalia) with the EC for a further 30 months of expanded activities. The new phase is in progress closely with the local farming communities, through providing training work (mostly safety training) which will be greatly expanded, and messages will be further spread by local radio and via Internet (there are cyber cafes in almost every Somali small town) and the schools. Moreover, different publications on crop pests and diseases of Somalia, IPM methodologies, including pragmatic information on crop protection principles were developed and some organic plant protection experiments are in progress.

Phytopathological analysis was developed with the aim to provide those involved in crop production and crop protection, in extension, research or training, with means to identify the most dangerous parasites' agents of disease on the main crops in Somalia. By field experiences it will be possible to record and list phytopathogenic microorganisms concerning the principle field, plantation and vegetable crops, and mineral and environmental disorders particularly spread.

As in other tropical areas, also in Somalia different pathogens (bacteria, fungi, viri) turned out to be able to cause dangerous diseases on several crops. While the description of the pathogens remains the same over time (even the scientific classification could be sometimes revisited), the disease strategies of prevention and control must be updated frequently taking in account the information coming from new innovative and effectiveness research characterised by a low environmental impact and the most appropriate useful cultural practices.

Considering the short life of agriculture in Somalia, intended as intensive field-work, able to offer new opportunities of job, to improve the local and national economies, as to point out the importance (nutritional, environmental, etc.) often for exclusive local crops, it was fundamental to work about the appropriate plant protection strategies of their crops using strategies without chemical impact, for a real benefit for the whole of Somalia.

In Somalia the farmers still cultivate their most important crops by traditional methods with low inputs and low yields. For example, the cultivation of legume crops is heavily influenced by climatic condition of the cultivated areas and by equipment availability. Most of the legume crops are developed along the Shabelle and Juba rivers. The rainfalls are concentrated in two periods per year; between April and June and in October-November while, most of the other months, are characterised by dry climatic conditions. During 2005, rains were below normal. June rains normally save the Shabelle and Bay regions from severe crops failure. Moreover, the coastal rains encouraged substantial planting thus ensuring an income source for riverine zone communities of the Shabelle and Juba Valley

Regions.

Similarly, sesame crop is very important for Somalian people. The sesame production in 2004 was of more than 10,000 MT. During 2004 rains were erratic in general and started late in most part of the country. In most of the southern and central regions, the rains were below normal, leading to crop failure and limited water availability. Rainfall was unevenly distributed over time and space across southern Somalia causing stress to both rain-fed and irrigated crops, like sesame, during April and May 2004. In addition to poor rainfall, an inefficient irrigation infrastructure and mismanagement of irrigation water sharing during 2004 have contributed to a poor production year. In any case sesame production allows different labour opportunities (land preparation, sowing, irrigation) giving livelihood options especially for the riverine zone communities. Moreover, by increasing sesame production further it will be possible to improve food security and reduce dependence on food imports. Under this cultivation in Somalia not a lot of pathogens are present, but few microorganisms (fungi) are able to determine remarkable damages if their attacks are heavy. Disease incidence on sesame crops in Somalia is particularly related to favourable climatic conditions (temperature and relative humidity, rainfall) and cultural practices (healthy soil, seed certification, cultivars' resistance, fertilization, irrigation, crop rotation, fungicide treatments) that need to be improved (Balestra 2006c).

10.2 Plant pathogens investigation

Among the crops spread most widely in Somalia, pulse crops, cucurbits, eggplant, onion, pepper, sweet potato, tomato, maize, sorghum, rice, cotton, sugar cane, sesame and sunflower turned out to be those most important as herbaceous plants while banana, citrus, mango, date palm and papaya are most important as arboreous ones.

On **pulse crops** (cowpea, chickpea, mungbean, pigeon bean, soybean), common fungal (downy mildew caused by *Pseudoperonospora cubensis*, Antrachnose, by *Colletotrichum*, Rust by *Uromyces vignae*), viral (*bean common mosaic virus strn blackeye cowpea* BLCMV, controlled by spray emulsions 1% of *Azadirachta indica*) and bacterial (bacterial leaf blight by *Xanthomonas axonopodis pv. phaseoli*) pathogens are mainly found. Their control can be affected by using disease-free seeds, disease-free planting materials, use of resistant cultivars, crop rotation and weed control. Moreover, proper fertilization, water management, plant spacing for air circulation and sunlight penetration within plants, clean farm tools after harvest, ploughing under all the plant debris, removal of infested plants and copper spray, together, give positive results. Inoculation of chickpea seeds with *T. harzianum* in a pot experiment reduced infection of seedlings from 70% in untreated controls to 18%. This fungus also gave effective control of charcoal rot when applied to seeds of groundnut, lentil, melon and *Vigna mungo*. The treatment increased germination and reduced post emergence mortality under pot culture conditions. Dry root rot of *Vigna radiata*, caused by *M. phaseolina*, was reduced by applying *T. viride* as a row treatment two days before sowing, under acid soil conditions (Balestra 2006b, 2006d; Emechebe & Florini 1997; Fernandez 1992; Oyekan & Naik 1987; Singh et al 1994; Tadesse 1995; Wydra & Rudolph 1999).

On **cucurbits**, antrachnose (by *Colletotrichum lagenarium*), powdery mildew (by *Erysiphe* spp., *Sphaerotheca fuliginea, Leveillula taurica*) and downy mildew (by *Pseudoperonospora* *cubensis*) resulted in a higher frequency of fungal pathogens. Prevention and organic control of these diseases is possible by use of disease-free seeds, anthracnose-resistant varieties, crop rotation, and regular application of copper compounds. Moreover, ploughing in or removing plant residues after harvest, spraying of sodium bicarbonate, pinolene (a pine resin derivative), mineral oil, monopotassium phosphate (potassium dihydrogen phosphate) and rapeseed oil (variety Canola), alone or in combinations, seem to be effective (Balestra 2006b; Jeffries et al 1990).

On **pepper plants** two main diseases were recorded. One is a fungal disease, cercospora leaf spot (caused by *Cercopsora capsici*). It causes damages at all growth stages with symptoms characterised by spots with grey-brown coloured centre. Its control can be obtained by using disease-free seeds, resistant cultivars, removing infected plants and by copper treatments. The other common disease on pepper is Cucumber Mosaic Virus (CMV). This virus has a wide host range and affects the pepper plants at all growth stages with symptoms that include curling and yellowing of the leaves, stunted growth, and malformed fruits. Its prevention/control is difficult and mainly related to the control of aphids that spread the disease, crop rotation and removal of weeds (Balestra 2006b; Reuveni et al 1998).

Tomato plants in Somalia turn out to be affected by different pathogens. Mainly these are alternaria blight (by *Alternaria alternata),* early blight (by *A. solani*), fusarium wilt (by *F. oxysporum* f. sp. *lycopersici*), late blight (by *Phytophthora infestans*), bacterial wilt (by *Ralstonia solnacearum*), bacterial speck (by *Pseudomonas syringae* pv. *tomato*), bacterial spot (*Xanthomonas vesicatoria*) and bacterial canker (*Clavibacter michiganensis* subsp. *michiganensis*). Seed selection and a warm treatment are suggested. In addition, the removal and destruction of infected leaves without putting them on the compost pile, the use of tomato, crop rotation, weed control, appropriate irrigation and fertilization and copper spray provide partial disease control (Balestra et al 1998; Balestra 2006b; Varvaro et al 2001).

On **maize plants** different fungal pathogens (Collar rot (*A. niger*), grey leaf spot (*Cercospora* spp.), sclerotium rot (*Corticium* spp.), corn rust (*Puccinia polysora*) are often recorded. The ergot (*Claviceps Africana*) and charcoal rot (*Macrophomina phaseolina*) diseases are typically observed on sorghum plants. On rice there can be found blast (*Pyricularia grisea, Magnaporthe grisea*) and rust (*Puccinia* spp.) fungal diseases. Their control is developed by using resistant cultivars, disease-free seeds, split application of nitrogenous fertilizer, proper water management and avoiding farm activities when plants are wet. Moreover, the elimination of diseased debris and crop rotation are strongly suggested (Balestra 2006b; Castellani & Mohamed 1981; Elad et al 1986; Gray et al 1990, 1991).

On **sesame plants** some fungal pathogens (*Alternaria sesami, Aspergillus niger, Corticium rolfsii, Fusarium oxysporum* f. sp. *vasinfectum, Macrophomina phaseolina*) are able to cause losses. Considering most of them as a predominant seed borne disease, the most important sanitary method is to ensure that the seeds are free of infection. After infection it is appropriate to remove infected plants and spraying the rest with Bordeaux mixture. Crop rotation has a strong influence on the survival of *C. rolfsii.* Several fungi have been investigated extensively as biological control agents for *C. rolfsii.* The mechanism of control was investigated by examining the hyphal interactions between *Trichoderma harzianum* and

C. rolfsii. Thin hyphal branches of *T. harzianum* grew towards and along those of *C. rolfsii*, forming several short branches and coils. Penetration sites on the host hyphae were observed after removal of *T. harzianum*. Control of root knot nematodes is affected by crop rotation to any crop other than cotton/legumes in reducing soil populations of these fungal pathogens. Moreover, seeds surface disinfection (0.5% sodium hypochlorite for 5 min and washing twice in sterile water) resulted useful (Abdou et al 1979; Balestra 2006b, 2006c; Berry 1960; Dolle & Hegde 1984; Elad et al 1986; Gray et al 1990, 1991; Mathur & Kabeere 1975; Ojiambo et al 2000).

On **cotton**, bacterial leaf blight (*Xanthomonas axonopodis pv. malvacearum*) and wilt (*Fusarium oxysporum* f. sp. *vasinfectum*) are common. About the fungal disease, soils in which Fusarium wilt occurs also favour root knot nematodes and these two parasites are often found together. Moreover, the nematodes also predispose the plant to fungal attack so that the control of nematodes is of major importance in reducing Fusarium wilt. Avoiding infested seeds, fields and the use of resistant varieties are suggested. This pathogen is seed-borne and in infected soil it is spread by boots, farm equipment and flood irrigation system. Rotation to any crop other than cotton reduces soil populations of the fungi and will reduce the severity of the disease in a subsequent cotton crop. However, the fungus will sustain itself on the roots of most plants including weeds (without causing any symptoms) and cannot be eliminated by crop rotation alone. Any field operation that moves soil from one location to another can spread spores of the fungus and introduce it to other fields. Cotton seed should never be produced in infested fields (Balestra 2006b; Ebbels 1975; Hillocks 1984).

Concerning **sugarcane plants**, alternaria leaf spot (caused by *Alternaria* spp.) is the most frequent one. The fungal pathogen causes damages to leaves, stem and flowers with symptoms that appear circular with lesions dark brown to black with concentric rings. Some lesions have distinct yellow halos on young plants. To prevent and to control this disease a strict sanitation of crop debris, the destruction of all infected plants, crop rotation, the use of disease-free seeds and a copper spray resulted effective only as preventive and if used in combination with cultural control strategies (Balestra 2006b).

Banana plants, resulted infected by antrachnose (by *Colletotrichum musae*) and leaf spot (by *Mycospherella musicola*) fungal diseases. To reduce the risk caused by their attack it is important to choose proper seeds and a selection of materials before planting: Sowing only disease-free seeds, transplanting only healthy seedlings, harvesting unripe but mature fruits are fundamental approaches. In addition, keeping the area free of weeds, removing and destroying infected parts but avoiding touching other plant parts, especially when these are wet and effect Bordeaux mixture spray, can help a lot to reduce disease infection and its spread (Balestra 2006b).

On **citrus trees**, scab fungal disease (by *Elsinoe fawcetti*) and Gummosis (*Phytophthora citrophthora*) resulted in spreading. Frequent copper applications on bark and foliage provide effects in reducing the development of these fungal diseases. Moreover, recently, symptoms resembling to those of Citrus canker (*Xanthomonas citri* subsp. *citri* ex *Xanthomonas axonopodis* pv. *citri*) were observed in southern Somalia. This pathogen is difficult to control, and further constant copper spray, if different studies in progress will confirm its presence, plant quarantine regulations and appropriate agronomical cultural practices against this pathogen should be carried on (Balestra 2006b).

On **date palm**, false smut fungal disease (caused by *Graphiola phoenics*) was observed. Its prevention and control could be carried on by the removal of infected branches and spraying the rest with Bordeaux mixture (Balestra 2006b).

Regarding **mango trees**, the antrachnose (by *Mangifera indica*) is the most widely spread disease, and its control is related to the removal of infected branches and spraying Bordeaux mixture on the rest (Balestra 2006b).

Papaya plants can be smut with the Ringspot virus (PRSV). Its symptoms may vary in intensity according to the age at which the plant becomes infected. Leaf symptoms are characterized by intense yellow mosaic and leaves distortion. The number of rings on the fruit can vary and the rings become less distinct as the fruit mature and yellow. Trees infected at a very young age never produce marketable fruit but rarely die. To prevent and control this virus it is suggested to use tolerant cultivars. Planting in partially isolated areas as far as possible from old infected papayas, check if diseased plants are present every 2 weeks, avoiding cucurbit plants in or near the orchard and systematic weed control reduce the vectors (aphid populations) (Balestra 2006b).

Apart from biotic stresses caused by different plant pathogens, in Somalia different abiotic stresses have been recorded. In particular, water stress, salt excess (in water and in soil), sunscald and different mineral (magnesium, potassium, calcium) deficiencies often determine favourable situations for plant pathogens attacks. To solve abiotic problems (deficiency/excess of mineral elements and/or unfavourable environmental conditions) on main crops in Somalia, it is important to consider each problem not simply in terms of the amounts of plant nutrients contained in the soil, but also taking into account other important factors like climatic parameters (ex.: temperature, light, water) to avoid stress situations for the plants. The presence of adequate quantities of plant nutrients in the soil is no guarantee if they will not be absorbed by the plant roots; mineral elements may be present in forms which are not available to the plants. For example, iron and zinc deficiency symptoms may be less severe under conditions of low light intensity, whilst boron deficiency effects are less severe and magnesium deficiency effects are more pronounced in wet seasons than in dry ones. Moreover, the rate of water absorption is less at lower temperatures than at higher ones and efficient intake is also dependent on good aeration. These facts may result in a water deficit within plants growing in wet soils when the air temperature is high. Always, when plants are grown in unsuitable environments, they react in specific ways. If light is insufficient, the green colouring matter of the leaf will be lacking and the leaves may be almost white (chlorotic); if the temperature is too high, the growth may be rank and soft; if water is insufficient, growth may be restricted, the tissues woody and the green of the leaves show a bluish tint. Moreover, deficiencies and excesses of the specific mineral elements produce characteristic effects on various organs of plants (seeds, roots, leaves, stems, buds, flowers, fruits) determining remarkable negative effects on the whole production.

10.3 First organic plant protection

Considering the vegetal species present/cultivated in Somalia, some of them were chosen and tested to promote organic plant protection strategies verifying their antimicrobial activities against plant pathogens. Recently in particular, the attention was focused on garlic (*Allium sativum*) and neem (*Azadirachta indica*) plants that are cultivated (garlic) and spread (neem) everywhere in Somalia (Pizzorno & Murraj 1996; Sridahr & Vijayalakshmi 2002).

The **Garlic plants** (Fam. *Liliaceae*) are native of Mediterranean regions of Europe and Africa with a Central Asian origin. The plant is cultivated everywhere around the world and is characterised by long narrow leaves that are flat like grass. Its bulbs are used as medicinal spices and are constituted by numerous bulblets or individual cloves grouped together and enclosed in a thin white skin. The first report about its use was found in Sanskrit documents (5,000 years ago) and in 1858 Louis Pasteur was the first scientist that pointed out its antibiotic properties. Its antimicrobial activity in medicine (with a broad spectrum activity against many bacteria, viruses and fungi) was verified during the last century. It is able to inhibit several dangerous bacterial (*Streptococcus, Bacillus, Brucella, Vibro* spp., as well as *E. coli, Proteus vulgaris, Salmonella enteritidis, Klebsiella pneumonia and Mycobacteria*), fungal (*Candida albicans* and *Cryptococcus neoformans*) and viral (*Herpes simplex* type 1 and 2, *Parainfluenza* virus type 3, *Human rhinovirus* type 2) human pathogens. Garlic extracts preparation was obtained blending 100 gr of grated and crushed garlic cloves. This extract plus 10 gr of soap (as emulsifier like potash-based one) were added to 10 l of water; all was well mixed and the mixture passed through a fine cloth. Later, the solution was diluted in 5 l of water and used immediately *in vitro* (spots test) and *in vivo* (spray on plants) tests.

Neem plants (Fam. *Meliaceae*) are known for their repellent, insecticidal, antibacterial and antifungal activities. Neem is a native plant from Central Asia; it is an evergreen tree characterised by leaves pinnate and serrate leaflets, with white flowers and fruits (drupes) ovoid yellowing coloured when ripe. Typically present in semi-arid, wet tropical and subtropical regions, this tree is present in all (south, centre, north) Somalian areas. It grows well also in dry/infertile areas. It is naturally distributed by different animals (ex: birds) and can be reproduced also from cuttings. Of this plant are mainly used leaves and seeds. No negative effect was recorded on humans, birds and other animals. In contrast, the active principle obtained from neem plants, the azadirachtin, is relatively harmless to butterflies, bees and ladybugs. Azadirachtin must be ingested to be effective so that pests which feed on plants are affected by its content. Moreover, constant spraying of flowering plants with highly concentrated neem products affect bees since they carry contaminated pollen and nectar to the hives. Neem leaf extraction was effected with 5 kg of seeds. Neem samples were collected from free-diseases plants, stored in dry conditions in air containers, away from direct sunlight and moisture. Attention was paid to avoid any direct contact with crude extract during processes of preparation and application. It was obtained by mortar and pestle, added in a pot containing 3 l of water and covered with a cotton cloth for 3 days. To utilise the extract in different tests, 1 l of neem seed extract was diluted with 9 l of water; 100 ml of soap were added (as emulsifier like potash-based one) and stirred well before using it on *in vitro* (spot tests) and on *in vivo* (spray on plants) tests. To obtain the best results, the appropriate neem extract colour (white-cream) was verified during its preparation, the soap was added before the water and then it was used immediately to avoid the floating of oil droplets.

Vegetal samplings (garlic and neem) were affected in Somalia in Hargeysa, Dooho, Arabsiyo and Gabilay areas. The potential antimicrobial activities of the vegetal extracts (bulbs and seeds, respectively) were tested by *in vitro* and *in vivo* experiments.

As first organic plant protection temptatives, the choice was restricted on some bacterial pathogens (bacterial speck, Pst, bacterial spot, Xv, and bacterial canker, Cmm) mostly dangerous and affecting tomato plants in Somalia. In addition, it was addressed due to the great importance of tomato crops, and due to typical climatic conditions recorded during recent years in Somalia. In fact, during 2005-2007 years, very heavy rain seasons after dry periods, seem a constant situation in most important rural Somalian areas. These phenomena seem to be in relation to the global environmental change and these factors, show the most remarkable negative influence/impact on disadvantaged rural areas greater than their impacts on the other human activities, especially in subtropical and tropical areas like those of Somalia. Considering these aspects, among plant pathogens, phytopathogenic bacteria seem to be those that found in such conditions the most suitable environmental/climatic conditions to attack, spread, survive and cause severe damages on several crops.

For *in vitro* tests, the spot technique was utilised. Fresh vegetal materials were washed under sterile distilled water and blotted with paper towels. Samples were then sliced into small pieces and blended using twister blender. The extract was obtained by centrifuging samples using sorvall RC % B (Newton, CT) at 8,000 g for 45 min to remove bigger parts. The extracts were collected and then kept at 4 °C until used within 24 hrs. Spot tests were carried on in lab and they were repeated 3 times with two replicates of each combination. Bacterial strains characterised by higher level of virulence of Pst, Xv and Cmm were utilised at 10^6 and 10^8 cfu/ml concentration. Natural extracts of garlic and neem plants were utilised at a concentration of 1%. Tests were effected on NSA medium (nutrient broth, 8 g/l; sucrose, 50 g/l, and agar 18 g/l). After distribution of each bacterial suspension (100 µl per Petri dish)/concentration, the vegetal extracts (4 drops, 30 µl each) were placed on and Petri dishes were incubated at 25 ± 2°C for 48-72 hrs; eventual inhibition halo's developed were observed and measured daily by a stereomicroscope.

To carry on *in vivo* tests, in greenhouse tomato plants 1 month old were used. Temperature (T) was maintained at 25 ± 2°C and at 15 ± 2°C for 12 hrs, respectively, and relative humidity (RH) was maintained between 70-80% during whole experiments. Each combination (bacterial strain/concentration/vegetal extract) was repeated 3 times and each *in vivo* experiment was 15 days long. In theses tests, the extracts of garlic and neem plants were used at a concentration of 1%. Bacterial strains where the same of those used in *in vitro* tests and were used at 10^6 cfu/ml concentration. For each combination (bacterial pathogen/concentration/vegetal extract) 80 tomato plants were used (60 to verify the effects of the vegetal extracts; 10 as control + only contaminated by bacterial pathogen; 10 as control – treated by copper oxhychloride 28%. Tomato plants were abundantly irrigated and each natural extract was sprayed until to obtain a uniform distribution on plants; 24 hrs later, bacterial suspensions were sprayed on tomato plants. Relative humidity was maintained at 90% (2 hrs before and 2 hrs after bacterial contamination) to favour stomata opening. After bacterial contaminations tomato plants were monitored daily for 15 days. Since the appearance of typical symptoms (6th day), disease incidence (number of necroses/diseased leaf area per cm^2 of tomato leaf were observed and analysed. *In vitro* and *in vivo* data were statistically analysed by a completely randomized design where all tests had three replicates. The mean separation was aided with Tukey's test ($p < 0.05$).

In vitro tests, garlic vegetal extract showed antibacterial effects respect to the Pst, Xv, Cmm tomato bacterial strains utilised. The garlic extract (1%) showed antibacterial activity on Pst, Xv and Cmm strains when the bacterial concentration was of 10^6 cfu/ml and also on Pst at 10^8 cfu/ml (Tab. 1). The best bacterial inhibition was obtained on Xv and Cmm strains when they were at concentration of 10^6 cfu/ml (Tab. 1).

The neem extract (1%) pointed out its antibacterial effect on all bacterial pathogens tested (Cmm, Pst, Xv). Its activity was even determined on plant bacterial pathogens when they were at 10^6 cfu/ml concentration (Tab. 1). By using the neem extract, the best inhibition results were obtained for Cmm pathogen (Tab. 1).

Tab. 1: *In vitro* antibacterial activity in spot agar tests by garlic and by neem extracts on *P. s.* pv. *tomato* (Pst), *X. vesicatoria* (Xv) and *C. m.* subsp. *michiganensis* (Cmm) on NSA medium after 72 h of incubation at 25 ± 2 °C. [1]Values expressed are mean ± S.D. of three experiments ($P \leq 0.05$). NID = No Inhibition Determined

Natural extracts	**Pst**		**Xv**		**Cmm**	
	cfu ml^{-1} 10^6	cfu ml^{-1} 10^8	cfu ml^{-1} 10^6	cfu ml^{-1} 10^8	cfu ml^{-1} 10^6	cfu ml^{-1} 10^8
Garlic (1%)	14 ± 1.3	13 ± 1.2	16 ± 1.3	NID	23 ± 1.1	NID
Neem (1%)	4 ± 0.8	-	4.5 ± 0.9	-	6.4 ± 0.8	-

Natural substances tested confirmed their antimicrobial effects respect to the Pst, Xv and Cmm reducing disease incidence (%) caused tomato bacterial pathogens by *in vivo* tests. By garlic extract, Pst, Xv and Cmm disease incidence was of 4% and 28%, 4% and 35%, 2% and 25% after 6 and 15 dd, respectively. By neem extract, Pst, Xv and Cmm disease incidence was of 9% and 40%, 10% and 45%, 4% and 41% after 6 and 15 dd, respectively (Tab. 2).

Tab. 2: Disease incidence (%) on tomato plants contaminated by *P. s.* pv. *tomato* (Pst), *X. vesicatoria* (Xv) and *C. m.* subsp. *michiganensis* (Cmm), (10^6 cfu/ml), after preventive leaf treatments by garlic and neem extracts (1%) ($p < 0.05$)

PST	Control	Garlic	Neem	Cu^{++}
Day 6	12	4	9	0.5
Day 9	48	17	35	3
Day 12	73	26	38	6
Day 15	86	28	40	8
Xv				
Day 6	14	4	10	0.7
Day 9	38	14	28	6
Day 12	65	25	43	10
Day 15	79	35	45	16
Cmm				
Day 6	8	2	4	0.2
Day 9	35	10	20	2
Day 12	60	16	37	4
Day 15	71	25	41	9

10.4 Discussion

In Africa the first biocontrol attempts were made in Kenya, against aphids in 1911. Later, different temptatives have been developed but the persistent use of chemical substances disturbed its spread and wide success. During last decades, to reduce spread of chemical substances and to preserve the environment as well as the human health, different organic approaches, especially against dangerous insects, were tested in different important rural African areas. Moreover, until now in Africa the use of microorganisms and natural substances to develop innovative biocontrol technologies as fundamental parts in biodiversity conservation have not been intensively studied to verify how they can be integrated into sustainable and suitable production systems.

Management of plant diseases by application of plant extracts has previously been suggested and the present results by extracts of garlic and neem plants were found to be effective in reducing the pathogens growth and the disease incidence caused by bacterial speck, bacterial spot and bacterial canker on tomato plants. Thus, in Somalia as well as in other African areas, organic control of plant pathogens shows a great potential and sustainable solution for the reduction of losses caused by dangerous phytopathogenic microorganisms.

Furthermore, human activities, a possible factor related to the outbreaks and the spread of different plant diseases, as those investigated here, seem to be related to global warming (Schaad 2006). An increase of different plant diseases in extreme conditions (extended hot and wet times, long and heavy periods of rains, storms) appears to favour phytopathogenic microorganisms like bacteria. These extreme climatic conditions help the CO_2-levels increase, and numerous studies provide solid evidence that global warming is still increasing at dangerous rate so that damages caused by these pathogens should be expected to increase.

During this project, among different organic plant protection techniques, the antimicrobial activity of extracts of some of the plant species commonly available in Somalia against main tomato bacterial pathogens was particularly evaluated. Extracts of *Allium sativum* and *Azadirachta indica* were found highly effective in inhibiting the growth of the phytopathogenic bacterial strains tested. No negative phytotoxic effect was recorded on tomato plants by using these natural substances. These results are in accordance with the previous findings showing the antibacterial activities of garlic and neem extracts in human health. In addition, aqueous extracts of leaves and fruits of *A. sativum* and *A. indica*, respectively, have been found to be effective against a wide range of pathogenic fungi, bacteria, and viruses. Thus, present results suggest that the fruit and seeds extracts here tested have remarkable inhibitory effect against *P. s* pv. *tomato, X. vesicatoria* and *C. m.* susp. *michiganensis* tomato bacterial pathogens.

Considering the losses caused by Pst, Xv and Cmm on tomato crops observed in Somalia, the results obtained by using native natural substances appear particularly interesting. Moreover, different agronomical practices (irrigation, fertilization) resulted fundamental to increase the control of plant pathogens like those studied on tomato crops and also the soil fertility as the climatic conditions should be taken in right account. Even the best results to reduce the disease incidence caused by Pst, Xv and Cmm bacterial strains, were obtained by using copper oxychloride, the natural extracts here tested remarked a great potential against these pathogens. The antibacterial effects of garlic and neem extracts resulted effectively at least for 10 days, point out their potential use in alternative to chemical compounds and particularly useful to protect tomato plants before bacterial arrival. It is important to remark that the antibacterial effects of garlic and neem extracts resulted active for Gram – (Pst, Xv) and Gram + (Cmm) bacterial microorganisms and, moreover, for bacterial pathogens causing parenchimatous (bacterial speck, bacterial spot) and vascular (bacterial canker) diseases. Further studies will be developed to evaluate the field doses of these natural substances to optimize their antimicrobial effects respect to Pst, Xv, Cmm and other plant pathogens in Organic Agriculture in Somalia. One side-effect is that the taste of garlic remains on sprayed plants one month after spraying, so it may be the best to avoid spraying near the harvest time; moreover, its effect is not selective. Thus, it is better to spray garlic extracts only as preventive strategy.

This can be considered to be the first report about a wide vision on plant diseases knowledge and organic plant protection strategies potentially developable in Somalia. During the next future Organic Agriculture and its useful plant protection strategies will spread more widely in different African countries. Of course, IPM organic plant protection approaches require the right political decisions/support and an open ecological vision by national and international contributors.

Organic Agriculture could be defined as the production of a system which promotes and enhances agroecosystem health, sustaining biodiversity, biological cycles, soil activity and cultural practices characterized by low power out-put. Consequently, our main objective is to promote IPM strategies that, integrated with local farmers' knowledges and native bioagents, will really improve plant protection techniques in Somalia and support their agriculture for better productions in quality and quantity.

10.5 Acknowledgements

The author wishes to thank all UNA's staff and in particular R. Del Cima, P. Caroli, I. Impiglia, D. Pedrazzoli, J.L. Aston, J. Backaert, A.J. Harberd, M.H. Ahmed "Suufi", F.W. Mohed, O.A. Jimale, M.M. Gaana, D. Chiavegatti, M.H. Egeh, A.A. Fidar and N.W. Schaad. Moreover, he would like to extend his particular thanks to all Somali people/farmers who offered their experience, knowledge, time, curiosity and patience allowing to focus on the main phytopathological problems in Somalia and who, inside a permanent unstable political situation, proved to be strongly motivated to give an opportunity to rebuild the lives of thousands unlike Somali people who believe in an agriculture based on their natural resources.

References

Abdou, Y.A.; El-Hassan, S.A. & Abbas, H.K. (1979) Seed transmission and pycnidial formation in sesame wilt disease caused by *Macrophomina phaseoli* (Maubl.) Ashby. Agricultural Research Review **57**(2), 63-69

Alasoardura, S.O. (1970) A checklist for fungi recorded from Nigeria. Occasional Publications of the University of Ibadan **1**, 35

Anaso, A.B. (1995) Effects of cultivars and seed treatment on sorghum diseases in Nigeria. Cereal Research Communications **23**(1-2), 153-159

Balestra, G.M. (2006a) Somalia Field People: 80 pp. Gangemi Ed

Balestra, G.M. (2006b) Crop Diseases of Somalia: 89 pp. UNA IPM Project

Balestra, G.M. (2006c) Sesame Diseases in Somalia: 25 pp. UNA IPM Project

Balestra, G.M. (2006d) Legumes Diseases in Somalia: 30 pp. UNA IPM Project

Balestra, G.M. & Del Cima, R. (2006) Mineral Deficiency Problems: 33 pp. UNA IPM Project

Balestra, GM and Varvaro L (1997) Influence of nitrogen fertilization on the colonization of olive phylloplane by *Pseudomonas syringae* subsp. *savastanoi*. Developments in Plant Pathology **9**, 88-92

Balestra, G.M.; Antonelli, M. & Varvaro, L. (1998) L. Effectiveness of natural products for *in vitro* and *in vivo* control of epiphytic populations of *Pseudomonas syringae* pv. *tomato* on tomato plants. Journal of Plant Pathology **80**, 251

Balestra, G.M.; Antonelli, M.; Fabi, A. & Varvaro, L. (1999) Effect of organic cropping system on the presence of *Pseudomonas syringae* pv. *tomato* in the phyllosphere and in the rhizosphere of tomato plants. Journal of Plant Pathology **81**, 227

Berry, S.Z. (1960) Comparison of cultural variants of *Alternaria sesami*. Phytopathology, **50**, 298-304

Castellani, E. & Mohamed, M.I. (1981) Peronospora disease of maize in Somalia. Rivista di Agricoltura Subtropicale e Tropicale, **75** (2/3), 223-230

Chattopadhyat, C. (1999) Yield loss attributable to Alternaria blight of sunflower (*Helianthus annuus* L.) in India and some potentially effective control measures. International Journal of Pest Management **45** (1), 15-21

Cheng, Y.H. & Tu, C.C. (1972) Effects of host variety, plant maturity, soil temperature, and soil moisture on the severity of Macrophomina stem rot of jute. Journal of Taiwan Agricultural Research **21**(4),273-279

Cloud, G.L. & Rupe, J.C. (1994) Influence of nitrogen, plant growth stage, and environment on charcoal rot of grain sorghum caused by *Macrophomina phaseolina* (Tassi) Goid. Plant and Soil **158** (2), 203-210

DeMooy, C.J. & Burke, D.W. (1990) External infection mechanism of hypocotyls and cotyledons of cowpea seedlings by *Macrophomina phaseolina*. Plant Disease **74** (9), 720

Dhingra, O.D. & Sinclair, J.B. (1977) An annotated bibliography of *Macrophomina phaseolina* 1905-1975. Brazil, Universidade Federal de Vicosa

Diourte, M.; Starr, J.L.; Jeger, M.J.; Stack, J.P. & Rosenow, D.T. (1995) Charcoal rot (*Macrophomina phaseolina*) resistance and the effects of water stress on disease development in sorghum. Plant Pathology **44** (1), 196-202

Dolle, U.V. & Hegde, R.K. (1984) Epidemiology and control of leaf blight of sesame caused by Alternaria sesami (Kawamura) Mohanty & Behera. Plant Pathology Newsletter **2** (1): 10

Ebbels, D.L. (1975) Fusarium wilt of cotton: a review, with special reference to Tanzania. Cotton Growing Review, **52** (4), 295-339

El Ahmad, M & Mouselli, M.N. (1990) Wilt and root rot of chickpea in southern Syria. Arab Journal of Plant Protection **8** (2), 60-67

El-Abyad, M.S.; El-Sayed, M.A.; El-Shanshoury, A.R. & El-Sabbagh, S.M. (1993) Towards the biological control of fungal and bacterial diseases of tomato using antagonistic *Streptomyces* spp. Plant and Soil **149** (2), 185-195

Elad, Y.; Zvieli, Y. & Chet, I. (1986) Biological control of *Macrophomina phaseolina* (Tassi) Goid by *Trichoderma harzianum*. Crop Protection **5** (4), 288-292

Else, J.P. (1995) Foliar and head diseases of sorghum. African Crop Science Journal, **3** (2), 185-189

Emechebe, A. & Florini, D.A. (1997) Shoot and pod diseases of cowpea induced by fungi and bacteria. In: Advances in cowpea research, edited by B.B. Singh, D.R. Mohan Raj, K.E. Dashiell, and L.E.N. Jackai. Copublication of the International Institute of Tropical Agriculture (IITA) and Japan International Research Center for Agricultural Sciences (JIRCAS). IITA, Ibadan, Nigeria, 176-192

Emechebe, A.M. (1975) Some aspects of crop diseases in Uganda. Kampala, Uganda: Makerere University, 43 pp

Fernandez, M.R. (1992) The effect of *Trichoderma harzianum* on fungal pathogens infesting soybean residues. Soil Biology and Biochemistry **24** (10), 1027-1029

Gangopadhyay, S.; Wyllie, T.D. & Teague, W.R. (1982) Effect of bulk density and moisture content of soil on the survival of *Macrophomina phaseolina*. Plant and Soil **68** (2), 241-247

Gordon, T.R. & Martyn, R.D. (1997) The evolutionary biology of *Fusarium oxysporum*. Annual Review of Phytopathology **35**, 111-128

Gray, F.A.; Kolp, B.J. & Mohamed, M.A. (1990) A disease survey of crops grown in the Bay Region of Somalia, East Africa. FAO Plant Protection Bulletin **38** (1), 39-47

Gray, F.A.; Mihail, J.D.; Lavigne, R.J. & Porter PM (1991) Incidence of charcoal rot of sorghum and soil populations of *Macrophomina phaseolina* associated with sorghum and native vegetation in Somalia. Mycopathologia **114** (3), 145-151

Hillocks, R.J. (1984) Production of cotton varieties with resistance to Fusarium wilt with special reference to Tanzania. Tropical Pest Management **30** (3), 234-246

Hillocks, R.J. & Bridge, J. (1992) The role of nematodes in Fusarium wilt of cotton in Tanzania. Afro-Asian Journal of Nematology **2** (1-2): 35-40

Jeffries, P.; Dodd, J.C.; Jeger, M.J. & Plumbley, R,A, (1990) The biology and control of *Colletotrichum* species on tropical fruit crops. Plant Pathology **39** (3), 343-366

Kannaiyan, J.; Greenberg, D.C.; Haciwa, H.C. & Mbewe MN (1987) Screening cowpea for resistance to major diseases in Zambia. Tropical Grain Legume Bulletin **34**; 23-26

Kovach, J.; Petzoldt, R. & Harman, G.E. (2000) Use of Honey Bees and Bumble Bees to Disseminate *Trichoderma harzianum* 1295-22 to Strawberries for Botrytis Control. Biological Control **18** (3), 235-242

Kushi, K.K. & Khare, M.N. (1979) Seed-borne fungi of sesame (*Sesamum indicum*) and their significance. Seed Research **7** (1), 48-53

Mandeel, Q. & Baker, R. (1991) Mechanisms involved in biological control of Fusarium wilt of cucumber with strains of nonpathogenic *Fusarium oxysporum*. Phytopathology **81** (4), 462-469

Manici, L.M. & Caputo, F. (1995) Biology and physiology of *Macrophomina phaseolina* from Italian sunflower-growing areas and possible methods of early selection for resistance to the pathogen. Sementi Elette **41** (3-4), 40-42

Martineau, J. (1994) *MSDS for Azatin-EC biological insecticide*. AgriDyne Technologies, Inc.

Mathur, S.B. & Kabeere F (1975) Seed-borne fungi of sesame in Uganda. Seed Science and Technology **3** (3/4), 655-660

Mguni, C.M. (1987) Diseases of crops in the semiarid areas of Zimbabwe: can they be controlled economically? In: Cropping in the Semiarid Areas of Zimbabwe. Workshop Proceedings, Harare. Agritex **2**, 417-433

Mihail, J.D. & Alcorn, S.M. (1984) Effects of soil solarization on *Macrophomina phaseolina* and *Sclerotium rolfsii*. Plant Disease **68** (2), 156-159

Morris, S.C. & Ward, K.R. (1985) Control of *Aspergillus niger*. CSIRO Division of Food Research Annual Report for 1984/5: 56

Musa, S.K.; Habish, H.A.; Abdalla, A.A. & Adlan, A.B. (1973) Problems of onion storage in the Sudan. Tropical Science **15** (4), 319-327

National Research Council (1992) Neem: A tree for solving global problems. National Academy Press. Washington, DC, USA

Nattrass, R.M. (1961) Host lists of Kenya fungi and bacteria. Mycological Papers **81**, 1-46

Nicholson, R.L. (1992) *Colletotrichum graminicola* and the anthracnose diseases of maize and sorghum. Bailey, J.A. & Jeger, M.J. (eds.) Colletotrichum: biology, pathology and control. Wallingford, UK; CABI International, 186-202

Ojiambo, P.S.; Ayiecho, P.O.; Narla, R.D. & Mibey, R.K. (2000) Tolerance level of *Alternaria sesami* and the effect of seed infection on yield of sesame in Kenia. Experimental Agriculture **36**, 335-342

Olaya, G. & Abawi, G.S. (1996) Effect of water potential on mycelial growth, and on production and germination of sclerotia of *Macrophomina phaseolina*. Plant Disease **80**, 1347-1350

Orzbay, N. & Newman, S.E. (2004) Biological control with *Trichoderma* spp. with emphasis on *T. harzianum*. Pakistan Journal of Biological Science **7** (4), 478-484

Oyekan, P.O. & Naik, D.M. (1987) Fungal and bacterial diseases of soybean in the tropics. Singh, S.R.; Rachie, K.O. & Dashiell, K.E. (eds). Soybeans for the Tropics: Research, Production and Utilization. John Wiley & Sons Ltd, 47-52

Özcan, M. & Boyraz, N. (2000) Antifungal properties of some herb decoctions. European Food Research and Technology **212** (1), 86-88

Pal, K.K.; Tilak, K.V.B.R.; Saxcna, A.K.; Dey, R. & Singh, C.S. (2001) Suppression of maize root diseases caused by *Macrophomina phaseolina*, *Fusarium moniliforme* and *Fusarium graminearum* by plant growth promoting rhizobacteria. Microbiological Research **156** (3), 209-223

Pande, S.; Harikrishnan, R.; Alegbejo, M.D.; Mughogho, L.K.; Karunakar, R.I. & Ajay, O. (1993) Prevalence of sorghum diseases in Nigeria. International Journal of Pest Management **39** (3), 297-303

Patel, K.K. & Patel, A.J. (1990) Control of charcoal-rot of sesamum. Indian Journal of Mycology and Plant Pathology **20** (1), 62-63

Pineda, J.: Nass, H. & Rodriguez, H. (1985) Effect of the inoculum density of *Macrophomina phaseoli* on infection of sesame seedlings. Agronomia Tropical **35**(4-6), 133-138

Pineda, J.B. & Avila, M.J. (1993) Losses caused by *Macrophomina phaseolina* on sunflower crop (*Helianthus annuus*). Agronomia Tropical Maracay **43**(5-6): 241-251

Pizzorno, J.E. & Murray, M.T. (1996) "*Allium Sativum*" A Textbook of Natural Medicine. Bothell, WA: Bastyr University Publications **1**

Plumb-Dhinsa, P. & Mondajne, A.M. (1983) Index of plant diseases and associated organisms of Mozambique. Tropical Pest Management **30**, 407-429

Porta Puglia, A. & Caratelli, A. (1982) *Aspergillus niger*, pathogenic on sorghum seed. Annali dell' Istituto Sperimentale per la Patologia Vegetale Roma **8**, 59-65

Prakash, A. & Rao, J. (1997) Botanical pesticides in agriculture, 35-103. CRC Press, USA

Quattrucci, A. & Balestra, G.M. (2005) Biological control of bacterial pathogens on tomato plants. 1st Int. Symp. on Biocontrol of Bacterial Plant Diseases, 20

Ragazzi, A.; Moricca, S.; Dellavalle, I. & Gonnelli, T. (1995) Behaviour of *Fusarium oxysporum* f. sp. *vasinfectum* isolates from Angola on callus tissue of cotton. Zeitschrift fur Pflanzenkrankheiten und Pflanzenschutz **102**, 493-501

Raupach, G.S.; Liu, L.; Murphy, J.F.; Tuzun, S. & Kloepper, J.W. (1996) Induced systemic resistance in cucumber and tomato against cucumber mosaic cucumovirus using plant growth-promoting rhizobacteria (PGPR). Plant Disease **80** (8), 891-894

Reuveni, R.; Dor, G. & Reuveni, M. (1998) Local and systemic control of powdery mildew (*Leveillula taurica*) on pepper plants by foliar spray of mono-potassium phosphate. Crop Protection **17** (9), 703-709

Ricaud, C.; Egan, B.T.; Gillaspie, A.G. Jr & Hughes, C.G. (1989) Sugarcane diseases and their world distribution. In: Diseases of Sugarcane, Major Diseases, 341-376

Rodrigues-Neto, J. & Malavolta-Junior, V.A. (1995) Diseases caused by bacteria on crucifers (Doencas causadas por bacterias em cruciferas). Informe Agropecuario, Belo Horizonte **17**, 56-59

Schaad, N.W. (2006) Emerging Plant Pathogenic bacteria and global warming, 51. 7th Int. Conf. On Pseudomonas syringae and related Pathogens, 12-16 Nov, Agadir, Morocco.

Schaad, N.W.; Postnikova, E.; Lacy, G.H.; Sechler, A.; Agarkova, I.; Stromberg, P.E.; Stromberg, V.K. & Vidaver, A.K. (2005) Reclassification of xanthomonas species pathogenic on citrus. Systemic Applied of Microbiology **28**, 494-518

Shen, C.Y. (1985) Integrated management of Fusarium and Verticillium wilts of cotton in China. Crop Protection **4** (3), 337-345

Singh, S.K.; Gupta, B.R. & Verma, V.S. (1994) Combined management of anthracnose and Cercospora leaf spots in mungbean. Indian Journal of Mycology and Plant Pathology **24**, 139-140

Sivan, A. & Chet, I. (1989) The possible role of competition between *Trichoderma harzianum* and *Fusarium oxysporum* on rhizosphere colonization. Phytopathology **79** (82), 198-203

Songa, W.; Ronno, W.K. & Danial, D.L. (1995) Production constraints of beans in the semi-arid eastern Kenya with special reference to charcoal rot. Breeding for disease resistance with emphasis on durability. Proceedings of a regional workshop for eastern, central and southern Africa, held at Njoro, Kenya, October 2-6, 1994, 251-255

Sridhar, S. & Vijayalakshmi, K. (2002) Neem: A user's manual. CIKS, Chennai

Stewart, R.B. & Yirgou, D. (1967) Index of Plant Diseases in Ethiopia. Haile Sellassie I University College of Agriculture, Experimental Station Bulletin, 30

Stromberg, P.E. & Vidaver, A.K. (2006) Amended classification of xanthomonad pathogens on citrus. Systematic Applied of Microbiology **29**, 690-695

Subrahmanyam, P. (1991) Control of seedling diseases of groundnut in Niger. Tropical Pest Management **37** (2), 118-119

Subrahmanyam, P.; McDonald, D. & Hammons, R.O. (1984) Rust. Porter, D.M.; Smith, D.H. & Rodriguez-Kabana, R. (eds.) Compendium of peanut diseases. St. Paul, MN, USA: American Phytopathological Society, 7-9

Swanepoel, A. (1990) Bacterial blight in cotton. Farming in South Africa, G.7, 1

Tadesse, N. (1995) ICC 6045. A kabuli chickpea, resistant to wilt and root rots in Ethiopia. International Chickpea and Pigeonpea Newsletter **2**, 17-19

Tarp, G.; Lange L. & Kongsdal, O. (1987) Seed-borne pathogens of major food crops is Mozambique. Seed Science and Technology **15** (3), 793-810

Quinn, J. & Myers, R. (2002) Cowpea: A versatile legume for hot, dry conditions. Thomas Jefferson Institute. http://www.jeffersoninstitute.org/pubs/cowpea_guide.pdf (accessed 2008-05-07)

Tu, C.C.; Cheny Y.H. & Hwang, S.C. (1991) Effects of solar energy and green manures on the control of southern blight of tomato. Plant Protection Bulletin (Taipei) **33** (1), 95-102

Van Emden, H.F.; Ball, S.L. & Rao, M.R. (1988) Pest, disease and weed problems in pea lentil faba bean and chickpea. Summerfield, R.J. (ed.), World Crops: Cool Season Food Legumes. Kluwer Academic Publishers. Dordrecht, The Netherlands, 519-534

Varvaro, L.; Antonelli, M.; Balestra, G.M.; Fabi, A. & Scermino, D. (2001) Control of phytopathogenic bacteria in organic agriculture: cases of study. Journal of Plant Pathology **83**, 244

Vos, J.G.M. & Frinking, H.D. (1997) Nitrogen fertilization as a component of integrated crop management of hot pepper (Capsicum spp.) under tropical lowland conditions. International Journal of Pest Management **43** (1), 1-10

Wada, A.C.; Bdliya, B.S.; Turaki, Z.G.S.; Imolehin, E.D. & Misari, S.M. (2002) Role of seed in disease transmission and the need for effective seed health testing centres in Nigeria. Outlook in Agriculture **31** (1), 55-62

Wood, C.M. & Ebbels, D.L. (1972) Host range and survival of *Fusarium oxysporum* f. sp, *vasinfectum* in N. W. Tanzania. Cotton Growing Review **49**, 79-82

Wydra, K. & Rudolph, K. (1999) Development and implementation of integrated control methods for major diseases of cassava and cowpea in West-Africa. Seifert, H.S.H.; Vlek, P.L.G. & Weidelt, H.-J. (eds.) Göttinger Beiträge zur Land- und Forstwirtschaft in den Tropen und Subtropen. Tropentag 1998. Stabilisierung und nachhaltige Entwicklung land- und forstwirtschaftlicher Systeme in den Tropen **133**, 174-180

Yassin, A.M. & Daffalla, G.A. (1982) A preliminary note on the present status of cotton wilt syndrome in the Sudan. Cotton et Fibres Tropicales **4**, 379-383

Youdeowei, A. (2002) Integrated pest management practices for the production of cereals and pulses. Cudjoe, A. & Braun, M. (eds.) Integrated Pest Management Extension Guide 2; Series, Ministry of Food and Agriculture, Ghana, GTZ and CTA

Yuan, H.X.; Li, H.L.; Wang, Z.Y.; Wang, S.Z. & Wang, Y. (1998) Studies on the control of cotton fusarium wilt by soil antagonists. Chinese Journal of Biological Control **14** (4), 156-158

Zahavi, T.; Cohen, L.; Weiss, B.; Schena, L.; Daus, A.; Kaplunov, T.; Zutkhi, J.; Ben-Arie, R. & Droby, S. (2000) Biological control of *Botrytis*, *Aspergillus* and *Rhizopus* rots on table and wine grapes in Israel. Postharvest Biology and Technology **20** (2), 115-124

Further websites (all accessed 2008-05-07)

APSNet. Citrus canker: The pathogen and its impact. http://www.apsnet.org/online/feature/citruscanker/ .

AVRDC. Cercospora leaf spot. http://www.avrdc.org/LC/pepper/cercospora.html .

AVRDC. Fusarium wilt. *Fusarium oxysporum f. sp. lycopersici.* http://www.avrdc.org/LC/tomato/fusarium.html.

AVRDC. Mungbean production. http://www.avrdc.org/LC/mungbean/tutorials.html .

CIMMYT. Pigeon pea. http://ppathw3.cals.cornell.edu/mba_project/CIEPCA/exmats/pigeonpea.pdf

Dr Giorgio M. Balestra is Plant Pathologist - Research Scientist and Lecturer of IPM and Organic Plant Protection courses at the Department of Plant Protection, Faculty of Agriculture, University of Tuscia, Viterbo, Italy. His research foci are biocontrol by natural substances and biocontrol agents, epidemiological studies and abiotic and biotic factors related to the diffusion and the survival of plant pathogens at rhizosphere and at phyllosphere level on plants of Mediterranean and Tropical-Subtropical areas.

Agricultural biodiversity is essential for adapting to climate change[1]

Johannes Kotschi

Johannesacker 6, 35041 Marburg, Germany, E-Mail: kotschi@t-online.de

11.1 Introduction

Agricultural biodiversity – shortly named agrobiodiversity - and climate change are rarely discussed in the same context, but there are close mutual links; agrobiodiversity is reduced through climate change and – at the same time – key to cope with the consequences of a changing climate. This interrelationship is particularly important as the entire agricultural diversity of genes, species and their agrarian ecosystems, represents the resource base for food. With climate change becoming reality, genetic resources are getting a new value as they are of vital importance for adaptation.

11.2 Effect of climate change on agriculture

The implications of climate change for agriculture have opened a new window in the discussion of agrobiodiversity. Today, the losses of agricultural diversity world wide are mainly caused by changing agricultural practices and production systems commonly named as the industrialisation of agriculture. Today, environmental change is a minor factor reducing the diversity of crops and livestock are likely to gain importance. Five climate change-related factors can be identified: the rise of temperature, changes in precipitation patterns, the rise of sea levels, increased incidence of extreme weather events and the increase of greenhouse gases in the atmosphere, of which carbon dioxide is the most prominent.

The rise in temperature - commonly known as global warming - is probably the most obvious phenomenon of climate change. Since 1861 the global mean annual temperature has increased by 0.6°C as atmospheric carbon dioxide concentrations have risen by 32%. It is unlikely that such a concentration was ever reached during the past 20 million years (Malhi et al. 2002). Emission scenarios suggest that we will have 550 ppm within the next 40 -100 years, almost double the pre-industrial concentrations. This rise will be accompanied by a further increase in temperature (IPCC 2001). Depending on the geographical region, scientific estimates propose an additional mean annual temperature rise of 1 - 5.8°C (Caldeira et al. 2003), and it is expected that the increases will be highest in tropical and subtropical regions. In principle, the speed of temperature rise will decide whether agro-ecosystems will be able to adapt. It seems likely, that it will be too fast for natural adaptation and without human support and will reduce species diversity; at best, organisms may survive through migration. In the affected regions, this will lower agricultural yields. A gloomy scenario is presented by Parmesan (cited in WWF 2003). It states that a temperature change of 2°C may result in a loss of some species but with possible management options at hand; with a rise of 4°C many species will be lost, with few coping options remaining.

[1] First published in GAIA 16/2, 98-101

Finally a rise of 6°C will leave no more potential to adapt to climate change. As an example, the impact of temperature on coffee production in Uganda is studied by Simonett (1999). He forecasts that an increase of 2°C will dramatically reduce the area suitable for Robusta coffee. Only higher-altitude areas would remain.

Indirect temperature effects may not be less important. The evaporation of soils is increased, decomposition of organic matter accelerated, the incidence of pests and diseases aggravated (Pimentel 1993).

Changes in water supply pose another problem of increasing importance. In the last century, subtropical regions were confronted most likely with around 3% less precipitation and more frequent droughts. Contrary to this, the northern hemisphere likely experienced 5 -10% higher rainfall (IPCC 2001). At the same time, increasing seasonal and regional rainfall irregularity has been observed and scientific research suggests that this trend will become more pronounced. An indication of this is the increasing use of drought-tolerant varieties at many tropical sites. Similar trends can be observed in animal husbandry. For instance, camels are replacing cattle and goats in extremely drought-prone areas of Ethiopia (Oxfam 2002).

Among the various greenhouse gases it is not only carbon dioxide that matters. Chlorofluorocarbons, for instance, have severely reduced the atmosphere's protective ozone layer. And, as a general rule, each one-per cent reduction causes a two-per cent increase in ultraviolet radiation reaching the earth. Research from Hawaii suggests that a 25% ozone depletion level can reduce soybean yield by 19-25% (Teramura et al. 1990). High levels of UV radiation may also nurture pests and diseases. For example, fungus infection rates in wheat increased by 9-20% when experimental UV radiation was increased by 8-16% above "normal" (Biggs & Webb 1986).

In summary, dramatic implications are expected for agriculture and food supply. But the global trend displays enormous regional differences. Ironically, the poorest are most at risk. Fischer et al. (2002) predict that by 2080 the 40 poorest countries, located predominantly in tropical Africa and Latin America, may lose 10-20% of their basic grain growing capacity due to drought. It is also argued that many rain-fed crops in the tropical belt of Africa and Latin America are already near their maximum temperature tolerance, and their yield may fall sharply with a further rise (Richards 2003). By contrast and for temperate regions, yield increases are expected due to higher temperatures, increased carbon dioxide levels and partly higher rainfall; a country like China could experience a rise in production by 25% (Fischer et al. 2002).

As a consequence, a two-pronged strategy is required: mitigation and adaptation. On the one hand, all possible efforts have to be made to reduce greenhouse gas emissions and mitigate climate change, in agriculture for instance by methods of Organic Farming (Kotschi & Müller-Sämann 2004). On the other, fast and appropriate action is needed to enhance capacity to adapt to changes already inherent in the system but not yet fully visible. Both, mitigation and adaptation, have to be part of anticipatory prevention strategies.

Box I: Pastoralists' innovative responses to drought

"For the third year in a row [1997-2000], the southern region of Ethiopia has received insufficient rainfall and endured a deadly drought. Most of the livestock - the source of livelihood for most of the people of this region - have died and all the vegetation has withered. For many in the Horn of Africa, this drought has put them on the edge of famine for months."

In 2000, the Oxfam partner, Action for Development, purchased 120 camels, which are more drought-resistant than cattle because they only need water every 10 days or so - to be used for hauling water. The introduction of camels for water transport has freed women, who used to travel 6-10 hours to bring back as much water as they could carry. Now they can devote themselves to other activities. Camels can also be used to plough the land when there is sufficient rainfall to attempt planting."

"Today, one camel can haul more than enough water for a family. The men have assumed the task of handling the camels, freeing the women to provide care for their families and return to a variety of income-earning activities."

"Adde Lokko Aaro, a mother of six children, lost more than four hundred goats and cattle in the drought and was on the verge of financial collapse. With three camels at the disposal of her village, Adde's responsibilities have been drastically altered: '[The camels] bring enough water for a number of households at a time,' Adde Lokko said. 'They [women] don't have to carry water on their backs, our men have started getting involved in the work of fetching water, which is normally the responsibility of women. We are pleased to witness that our camels have shared our burden.'"

Cited from: Oxfam (2002)

11.3 Agrobiodiversity – how much shall be conserved?

The continuous loss of agricultural genetic resources is a matter of growing concern - a process in which climate change is gaining importance (IPCC 2002). In search of adequate responses, one has to consider that genetic resources are not only a victim; more importantly, they are crucial to coping with climate change (Umweltbundesamt 2004). Plants and animals with no economic value so far may become important, and new breeds able to tolerate environmental stress will be needed. But how much agrobiodiversity should be conserved to secure our future? Can we base public conservation strategies on mathematical modelling which sets priorities for conservation and the optimal degree of conservation is calculated as proposed for instance by Weitzman (1993, 1998)? Or must we conserve all we have because the future needs for human survival are unknown?

Valuing genetic resources with mathematic models is not yet relevant to decision-making. On the other hand it seems to be unrealistic to conserve all of them irrespective of any valuation. But as a basic principle it is evident that a maximum of genetic resources has to be conserved at the lowest possible public cost. Hence a conservation concept is required that goes far beyond *ex-situ* conservation, being the predominant approach for plant genetic resources. Storage of seeds in refrigerated banks or botanical gardens is essential. But this method exceeds the capacity of public funding, is of limited scope and of limited security (FAO 1997, GRAIN 1996). Therefore, gene banks can only be complementary to a more comprehensive conservation approach (Hammer 2004).

Such a more comprehensive approach relies primarily on *in-situ* concepts, managed by farmers and farming communities doing conservation and breeding on their farms and in their villages. Farmers have done so over thousands of years, have been ignored or neglected by the formal seed sector during the past 40 years and, since recently, are slowly being rehabilitated. On-farm conservation is not necessarily less costly, but the costs are mainly borne by farmers whereas the benefits are private and public.

Latest concepts of *in-situ* conservation follow the idea that conservation and use of genetic resources are closely linked. True to the slogan "use it or lose it", plant species or animal breeds should be used whenever possible, should contribute to securing rural livelihoods and to rural culture. "As long as farmers themselves find it in their own best interest to grow genetically diverse crops, both farmers and society as a whole will benefit at no extra cost to anyone" (Bragdon et al. 2004).

As a consequence, economic or social benefits have to be found for seemingly useless crops or farming systems and value has to be discovered in them. Some examples of adding economic value are: wild plants may be used for medicinal purposes, wheat landraces grown under Organic Agriculture may get a higher price, farming communities as a whole may profit from agro tourism if they maintain their diversity, etc. However, it will not be possible to find a market for everything that should be protected. Therefore, a remainder will have to be protected without "using" it - a service that has to be paid for by the public.

11.4 Species adaptation – but how?

Another argument that calls for a revised understanding of agrobiodiversity conservation is adaptation to climate change. This is because the ecosystem's capacity to be resilient and adaptive to environmental change relies fundamentally on genetic diversity. Adaptation describes a plant's, animal's or ecosystem's capacity to adjust to changes such as heat, drought, or salinity – an adjustment enabling them to overcome constraints, take advantage of new opportunities and cope with the consequences of changing environments.

Adaptation is a process. What matters is not so much the drought-resistant minor millet landrace, well stored in isolation and deep-frozen in a gene bank, but rather, exposure to the environment, on farmers' fields and considering the wide agro-ecological variations of sites. Resistance of plants to environmental stress (e.g. drought tolerance) is mostly a multi-genetic characteristic best developed by *in-situ* exposure to it. In contrast, it is difficult to achieve such traits through genetic engineering.

The social dimension is no less important. Adaptive capacity building has to address the poor and should enhance their human and social capital. The focus on women addresses the fact that in rural societies everywhere women have always been the seed keepers, the preservers of genetic resources.

Such a strategy as outlined above addresses regional and local agro-ecological variations. It offers site-specific solutions contrasting with those of the corporate sector that follows the law of economy of scales and aims to distribute a standardised variety or a whole cropping system technology as widely as possible.

Box II: Minor millets save the poor from starvation

"Sankappa is a small farmer owning three hectares of dry land in Vittalpura village of Bellary district in Northern Karnataka, India. This village is situated in the semi-arid Deccan Plateau and receives annual rainfall of 500 mm in two to three months a year, which allows one crop during July to October. Sankappa like his forefathers and other farmers of the village is growing foxtail millet. (...) The amount of rainfall during the last four years continuously dropped during the last four years in this part of the country. It was below 300 mm in 2003. 'All other crops failed due to extreme drought, and my family and livestock were saved from starvation by the harvest from foxtail millet,' says Sankappa. The (...) varieties grown and conserved by the villagers have excellent drought resistance."

"Eight minor millet crops grown in different regions of Africa, Asia and Eurasia are finger millet (*Eleusine coracana)*, proso millet (*Panicum miliaceum*), little millet (*Panicum sumatrense*), barnyard millet (*Echinochloa crusgalli* and *E. colona*), kodo millet (*Paspalum scrobiculatum*), teff (*Eragrostis tef*) and fonio (*Digitaria iburua*). Little millet and kodo millet were domesticated in India."

"The long history of minor millet cultivation and its spread to different regions of the world that are notable for extremely harsh farming conditions generated considerable genetic variability in these crops. (...) Global neglect of the minor millets and increasing emphasis on few elite food crop species are precariously narrowing the food security basket. The most disadvantaged by this food production policy are the poorest of the poor (...) The shrinking number of food crops in the regional and global food basket is restricting the opportunity of farmers in difficult regions to use their land resources, environment and traditional knowledge."

Cited from: Bala Ravi (2004).

11.5 Urgent action is required

There is little awareness among the various international development initiatives of the close relationship between climate change and food security and the role agrobiodiversity has to play. This concerns without distinction the programmes to fulfil the Millennium Development Goals (MDGs), the National Adaptation Plans for Action (NAPAs) by the United Nations Framework Convention on Climate Change (UNFCCC, UNEP 2005) and others. Adaptation to climate change in agriculture - if discussed at all - deals mainly with improved water management (in view of more frequent drought and flooding events). Agrobiodiversity - although being a fundamental resource for adaptation - is almost forgotten.

Instead, it must become imperative to manage agrobiodiversity in a sustainable way and to use it systematically to cope with the coming environmental challenges. The following aspects deserve consideration:

- Stronger coordination is needed between main global programmes such as the United Nations Framework Convention on Climate Change, the Convention of Biodiversity and the International Treaty on Plant Genetic Resources for Food and Agriculture (CBD 2003 and UNEP 2005).

- Agrobiodiversity conservation is to be made a basic component of adaptation strategies to climate change.
- Programmes that manage agricultural genetic resources require re-orientation in their strategies. Formal institutional systems based on gene banks (*ex-situ* conservation) must be broadened to an integrated management system that includes the farmer-based (*in-situ*) conservation (Almekinders 2003).
- *In-situ* conservation of agricultural biodiversity must be made an integral part of agricultural development and be supplemented by *ex-situ* conservation.

Only the public sector can take the lead in implementing such a comprehensive approach, in which the private sector has an important supportive role. National and intergovernmental laws and regulations will have to provide the necessary legal frame.

The role of civil society organisations (CSOs) as well as the corporate sector has become increasingly important to fill this frame with development reality on the ground. So far, CSOs have taken a front-runner position in developing and spreading suitable concepts at grassroots level. For instance:

- They have catalysed worldwide a boom of farmer initiatives that practice Organic Agriculture (Kotschi 2005) based on maintaining biodiversity, avoiding the use of hybrid seeds and prohibiting transgenic crops.

- They are increasingly supporting local seed conservation initiatives, such as Navdanya in India, that aims to "empower local communities (...) to protect and conserve their biodiversity and defend their community rights to seeds and knowledge" (Navdanya 2000).

- They have founded an alternative market for plant breeding and seed production. Mainly in Europe, various initiatives have emerged that maintain, improve and make available open-pollinating varieties of cereals and vegetables (Almekinders 2002), many of which are the result of crossbreeding and selection over centuries and in danger of getting lost. Kultursaat eV, an association in Germany that releases new varieties as public property, is a prominent example (Kultursaat 2007).

All such activities make it very clear: genetic resources must remain largely in the public domain with well-balanced benefit-sharing concepts among the various stakeholders that use and conserve agro-genetic resources. Generally, the *in-situ* approach offers a great chance to shape a future worth living.

11.6 Acknowledgement

This article was written with support through a grant from the sectoral project "People, Food and Biodiversity" of the Deutsche Gesellschaft für Technische Zusammenarbeit (GTZ) and is part of its awareness-raising activities.

References

Almekinders, C. & Jongerden, J. (2002) On visions and new approaches. Case studies of organisational forms in organic plant breeding and seed production. Working Paper Technology and Agrarian Development, Wageningen University.

Almekinders, C. (2003) Institutional changes for integrated management of agricultural biodiversity. International Potato Center (Ed.) CIP-UPWARD, Conservation and sustainable use of agricultural biodiversity: a sourcebook. Users' Perspectives with Agricultural Research and Development, Los Banos, Laguna, Philippines, 571-577

Bala Ravi, S. (2004) Neglected millets that save the poor from starvation. LEISA India **6** (1), 34-36

Biggs, R.H. & Webb, P.G. (1986). Effects of enhanced UV-B radiation on yield, and disease Incidence and severity for wheat under field conditions. Ecological Sciences, NATO ASI Series G-8

Bragdon, S (Ed.) (2004) International law of relevance to plant genetic resources: a practical review for scientists and other professionals working with plant genetic resources. IPGRI, Rome, Issues in Genetic Resources **10**, 93

Caldeira, K.; Jain, A.K. & Hoffert, M.I. (2003). Climate sensitivity uncertainty and the need for energy without CO_2 emission. Science **299**, 2052-2054

CBD (2003) Interlinkages between biological diversity and climate change. CBD Technical Series No 10. Secretariat of the Convention on Biological Diversity. Montreal

FAO (1997) The State of the World's Plant Genetic Resources. Rome, FAO

Fischer, G.; Shah, M. & Veldhuisen, H.v. (2002) Climate change and agricultural vulnerability. International Institute for Applied Systems Analysis. Report prepared under UN Institutional Contract Agreement 1113 for World Summit on Sustainable Development. Laxenburg. Austria.

GRAIN (1996) Ex-situ conservation: from the field to the fridge. Seedling. June 1996

Hammer, K. (2004) Resolving the challenge posed by agrobiodiversity and plant genetic resources - an attempt. J. Agr. Rur. Dev. Tropics Subtropics, Beih. 76, 184 pp

IPCC (2001) Climate change 2001: the scientific basis. Cambridge, Cambridge University Press

IPCC (2002) Climate change and biodiversity. Technical Paper V. WMO and UNEP

Kotschi, J. & Müller-Sämann, K. (2004) The role of Organic Agriculture in mitigating climate change – a scoping study. Bonn, International Federation of Organic Agricultural Movements

Kotschi, J. (2005) Mehr Ökologie – weniger Hunger? Ernährungssicherung und ökologische Landwirtschaft. Politische Ökonomie **90**, 59-61

Kultursaat (2007) Der Verein. www.kultursaat.com/start.html (accessed April 20, 2007)

Malhi, Y.; Meir, P. & Brown, S. (2002) Forests, carbon and global climate. Swingland, I.R. (Ed.) Capturing carbon and conserving biodiversity - The market approach. London: Earthscan, 15-41

Navdanya (2000) Biodiversity conservation, Sustainable Agriculture and food security. Proposal for Navdanya Core Programme. Navdanya, India

Oxfam (2002) Drought relief in southern Ethiopia. Sustainable water supply and food security systems established by Oxfam partner Action For Development (AFD) has helped hundreds of communities in southern Ethiopia survive severe drought. www.oxfamamerica.org. (accessed 1 June 2006)

Pimentel, D. (1993) Climate changes and food supply. Forum for Applied Research and Public Policy **8** (4), 54-60

Richards, M. (2003) Poverty reduction, equity and climate change: global governance synergies or contradictions? Globalisation and Poverty Programme. ODI. London

Simonett, O. (1999) Potential impacts of global warming, Case Studies on climate change, GRID-Geneva, Geneva

Teramura A.H.; Sullivan, J.H. & Lydon, J. (1990) Effects of solar UV-B radiation on soybean yield and seed quality: a six-year field study. Physiologia Plantarum **80**, 5-11

Umweltbundesamt (Ed.) (2004) Integration of biodiversity concerns into climate change mitigation. A toolkit. Umweltbundesamt, Berlin

UNEP (2005) Integration of biodiversity considerations in the implementation of adaptation activities to climate change at the local, sub national, national, sub regional and international levels. Note by the Executive Secretary, UNEP/CBD/AHTEG-BDACC/1/2 4 July 2005 www.biodiv.org/doc/meetings/cc/tegcc-04/official/tegcc-04-02-en.pdf (accessed 5 May 2007)

Weitzman, M. (1993) What to preserve? An application of diversity theory to Crane Conservation. Quarterly Journal of Economics **108**, 157-183

Weitzman, M. L. (1998) The Noah's Ark problem. Econometrica **66**, 1279-1298

WWF (2003) A user's manual for building resistance and resilience to climate change in natural systems. Climate Change Programme. Berlin: World Wild Life Fund

More ecology – less hunger?

Johannes Kotschi

Johannesacker 6, 35041 Marburg, Germany, E-Mail: kotschi@t-online.de

12.1 Introduction

"Organic Agriculture cannot feed the world" is a statement often heard in research and development circles. The argument is that, without chemical inputs and without genetic engineering, hunger cannot be alleviated. At the same time, the statement suggests that conventional agriculture can solve the problem. The truth, however, is much more complex than this.

12.2 Who suffers from hunger?

Two facts demand to be addressed when considering the problem of hunger. Firstly, according to the latest findings, 80% of the hungry live in rural areas and two thirds of them are smallholders (Task Force on Hunger, 2004). Secondly, in purely quantitative terms, the world produces enough food to provide everyone with enough to eat. Thus, the reason for hunger does not lie in the scarcity of food on the planet. People suffer from hunger because they are poor and because global resources are not equally distributed. Many people do not have access to sufficient land and other resources necessary to be able to grow their own food. Trade regulations and unattractive terms of trade discourage smallholders from producing a surplus for the market and discourage traders from serving marginal areas. The phenomenon of hunger is man-made. Children, women and men are hungry because they cannot afford to buy food or because sufficient food is not available in local markets.

If it is true that hunger primarily affects people in rural areas and that it originates from a lack of purchasing power and the unequal distribution of resources, then development strategies have to be aimed at empowering the hungry to produce their own food. Here, Organic Agriculture plays a key role and should be part of an appropriate strategy to combat hunger.

12.3 Why Organic Agriculture?

Organic Agriculture is commonly equated with "no mineral fertiliser and no pesticides". This is not, however, a sufficient description and reflects a false understanding. The guiding principle of Organic Agriculture is the cultivation of land or raise of animals in harmony with nature or – expressed in modern terminology – according to the laws of ecosystems. Natural processes are enhanced; nutrient and energy flows are kept as cyclic as possible; crop and animal husbandry are closely linked. The agricultural enterprise – "the farm" – with its people, land, plants and animals is conceived as a manifold whole, as an organism (Koepf et al., 1976).

Organic agricultural methods aim to use nutrients as efficiently as possible and not only once; they are reused as often as possible through largely closed nutrient cycles. Mineral fertiliser is not excluded but it is strictly limited. Fertilisers are applied primarily in order to enhance soil fertility (Pfeiffer 1947), especially in marginal areas. In contrast, in conventional agriculture, fertilisers serve primarily to maximise yields. Optimising inputs instead of maximising outputs (Kotschi 2000) is probably the most profound difference between the two types of agriculture.

Thus the key issue is to make optimum use of resources for production in marginal areas where resources are, by definition, limited. Globally, there are large areas and a huge group of people that could benefit from this. Resource-poor farmers need to be empowered to make better use of their minimal inputs – nutrients, water and seed – to obtain at least modest yields so that they can secure their subsistence and, wherever possible, sell their small surpluses.

12.4 Can Organic Agriculture generate adequate yields?

Research provides little evidence about the performance of Organic Agriculture in resource-poor areas. However, the boom in farmer initiatives worldwide that seek to work with ecologically-oriented agricultural techniques is an indication of the widespread conviction that this is a promising approach (Parrot & Marsden, 2002, Pretty et al., 2003). These initiatives by smallholder farmers – some of them supported by international donors in rural development, some of them genuinely local initiatives – have been very successful. For example, in the northern Indian foothills of the Himalayas, many villages have converted to ecological agriculture. After years of input-intensive farming under the banner of the "Green Revolution" it has been possible to rehabilitate the degraded soils and, within a relatively short period of time, the villagers have achieved good yields that are, in part, higher than would have been the case had they continued with conventional methods of production (Kotschi, 2003).

Without doubt, under very fertile conditions, Organic Agriculture cannot bring the same physical yields as can conventional agriculture. On degraded or marginal soils, however, such success stories are numerous and demonstrate that Organic Agriculture can compete successfully with conventional agriculture. In economic terms, the superiority of Organic Agriculture is sometimes even more pronounced. The continuous deterioration in terms of trade – increasing costs of production versus stagnating or decreasing prices of products – supports this trend (Kotschi, 2000). However, the most important argument in favour of Organic Agriculture as an approach to alleviating hunger is that it is less susceptible to production risks in periods of drought – an impact that has been obvious during the past three years in the poorer areas of southern Africa.

12.5 A global strategy

Organic Agriculture is more than just a technology. In concert with obligatory standards for production and processing and a well-functioning system of inspection and certification, it has become a comprehensive strategy of global dimensions. Be it green tea from China, coffee from Mexico or cotton from Tanzania, organic foods and textiles are being produced worldwide for the wealthy North: for Europe, North America and Japan. The binding character and the transparency of the strategy have created confidence among consumers. For 2006, the trade volume of organic products was estimated at 40 billion USD (Willer & Yussefi, 2007) and there are annual increases of more than 10%.

Standards of production and processing are the backbone of this strategy. They serve to increase consumer confidence and they stimulate demand for organic products.

At the same time, producers are obliged to seek technological alternatives to be able to meet the standards. This makes Organic Agriculture a generator of innovations, many of which have already been integrated into mainstream agriculture. Farmer-led research on biological pest control in fruit cultivation (apples and pears) is a good example of this from Europe.

Organic Agriculture is increasingly promoted by governments. The EC Organic Farming Regulation, which entered into force in 1991, was a milestone in this development (Schmidt & Haccius, 1998). By 2007, more than 60 states had regulations on Organic Agriculture; a further 18 states were in the process of drawing up such rules (Willer & Yussefi 2007). Private sector, internally binding guidelines have given way to national and international regulations (the latter including the Codex Alimentarius). "Organic" is no longer defined by farmers but by governments through laws and regulations. This can be seen as a great success and as an indicator for public recognition and even mainstreaming. But does Organic Agriculture contribute to poverty alleviation and food security?

12.6 Food security?

The boom in ecologically-oriented smallholder initiatives in developing countries is evidence that Organic Farming is superior to conventional approaches, particularly in poverty-stricken areas. But the organic market is closed to most producer groups. Many of them are unable to meet the requirements of international standards. An example may illustrate this.

In many parts of Asia, pesticide residues in vegetables have reached such an alarmingly high level that eating the vegetables jeopardises human health. Therefore, in North Sumatra, Indonesia, local non-governmental organisations (NGOs) and the farmer groups they support have developed alternatives to chemical pesticides. They have also begun defining their own standards and local guarantee systems so that they can market their pesticide-free produce locally (Kotschi 2004). The demand for such vegetables appears to be enormous – both in the populous towns of North Sumatra and in neighbouring Singapore. A closer look at their standards reveals that the use of synthetic mineral fertilisers is reduced but not completely excluded. This is primarily because there is not enough organic fertiliser available locally. During the "Green Revolution", the water buffaloes were replaced by tractors and mineral fertilisers – a "development" that is not easy to reverse. Having chosen to convert to organic, the smallholders now find that their use of even small amounts of synthetic fertilisers would violate the International Basic Standards and the Indonesian National Standards.

This reveals a general dilemma involving standards. That they are needed and beneficial is indisputable. Yet, at the same time, they are increasingly becoming a hindrance to expansion of Organic Agriculture. As is the way with most regulations, they have become more and more elaborate, sophisticated and expensive to implement; their development has taken on a life of its own. Moreover, the intention to harmonise different standards while at the same time making them more detailed stands in contradiction to the basic principle of Organic Agriculture: the "self-regulating farm organism" that needs to be free to shape its own course and to act on its own responsibility if it is to develop in the way best suited to its location-specific conditions. Present-day standards increasingly restrict this scope for development.

Who benefits from this? It is doubtful that the market really profits from this trend, since the costs of regulation are considerable. This process has not only reduced the scope of Organic Agriculture as a generator of innovation; it also excludes the more resource-poor farmers.

12.7 Poverty alleviation?

There are other groups of smallholders who, while they can comply with the standards, are still unable to access the international market:

- The quantities they produce are often too small and not homogenous in quality.
- They offer products that are not in demand on the international market or that are too readily perishable (e.g. vegetables). They also have little knowledge of marketing partners and marketing routes.
- Organic certification is often too expensive and complicated; this applies even to the cheaper scheme of group certification for smallholders who market their produce jointly.

Even more important, however, is the fact that demand for organic products in the industrialised countries is rising significantly more slowly than is the supply from the developing countries. For some products, such as coffee, saturation point has already been reached.

This means, therefore, that the current system of certified Organic Agriculture is not a route to food security or poverty reduction, since smallholders in the developing world are to a large extent excluded from the international market. In their efforts to farm more organically, these smallholders can expect no support (for example in the form of advice) from the international market. There is therefore a need to develop national, local and regional markets (Flörchinger et al. 2007). One can envisage such markets being governed ideally by a diversity of standards and guarantee systems arising from the relevant natural, cultural and socio-economic contexts rather than being imposed "from outside".

12.8 Conclusions

The past thirty years of Organic Agriculture have been a tremendous success. But with the present concepts and strategies, expansion alone will not be sufficient to master future challenges. The time has come to consider a reorientation.

(1) In resource-poor areas where food security is the main concern, Organic Agriculture is technologically clearly superior to conventional approaches. But there is a tremendous need for training and advice.

(2) Standards are undoubtedly necessary, and an organic market without them is inconceivable. But instead of becoming more detailed, international standards should be significantly simplified and should be couched in more general terms. Only then does it make sense – given the great diversity of site conditions – to strive for equivalence.

(3) We must rediscover values rather than standards as the subject of our discourse. The principles and models of Organic Farming need to be developed and applied more emphatically as a reference point.

(4) The importance of the global market for the development of smallholder agriculture is overestimated. The international market takes little account of objectives such as food security and can even enhance poverty. Instead, local and regional markets have to be supported.

(5) We should remember the pioneering role of Organic Agriculture, which constantly produces innovations for the mainstream. This innovation development should focus on resource-poor regions, on food security and on poverty alleviation. The main areas of research and development are: a) use of renewable energies in agriculture, b) building up regional markets with simple certification systems, c) adaptation to climate change and d) intensification of cropping and livestock systems through better use of plant and animal genetic resources (biodiversity).

A reorientation of this sort will need all the strength we can muster. It is necessary if we are to keep pace with the challenges of a rapidly changing world.

References

Flörchinger, F.; Bernd, A.; Becker, T.; Schrimpf, B. & Kotschi, J. (2007) Local marketing of organic products in developing countries. Guidelines for practitionners. AGRECOL. Göttingen. www.agrecol.de/index.php?option=com_content&task=view&id=65&Itemid=76 (accessed 2008-05-14)

Koepf, H.H.; Petterson B.D. & Schaumann W. (1976) Bio-dynamic agriculture, an introduction. The Anthroposophic Press, Spring Valley NY

Kotschi, J. (2000) Poverty alleviation in the South – can Organic Farming help? Alföldi T.W.; Lockeretz, W. & Niggli, U. (2000) Proceedings 13th International IFOAM Scientific Conference, Basel, 652-655

Kotschi, J. (2003) Biodiversity Conservation, Sustainable Agriculture and Food Security. Navdanya / India. Evaluation Report. Marburg

Kotschi, J. (2004) Agriculture in harmony with nature – A joint initiative in North Sumatra, Indonesia. Consultancy Report for Brot für die Welt and EED Germany

Parrot, N. & Marsden, T. (2002) The Real Green Revolution, Organic and agroecological farming in the South. Greenpeace Environmental Trust, London

Pfeiffer, E. (1947) Soil fertility, Renewal & Preservation. Faber & Faber, London

Pretty. J.; Morison, J.I.L. & Hine, R.E. (2003) Reducing food poverty by increasing agricultural sustainability in developing countries. Agric. Ecosys. Environ. **95**(1), 217-234

Schmidt, H.P. & Haccius, M. (1998) EU Regulation „Organic Farming“ – A Legal and Agro-Ecological Commentary on the EU's Council Regulation (EEC) No 2092/91

Task Force on Hunger (2004) Halving hunger by 2015: A framework for action. Interim Report, Millennium Project. United Nations, New York

Willer, H. & Yussefi, M. (2007) The world of Organic Agriculture 2004 – Statistics and emerging trends. IFOAM and FIBL, Bonn and Frick